ON THE ROOF OF THE ROCKIES

Books by

LEWIS R. FREEMAN

In the Tracks of the Trades

Down the Columbia

Down the Yellowstone

The Colorado River

Down the Grand Canyon

On the Roof of the Rockies

ON THE ROOF OF THE ROCKIES
The Great Columbia Icefield of the Canadian Rockies

Lewis R. Freeman

Photos by Byron Harmon

Foreword by Emerson Sanford and Janice Sanford Beck

VANCOUVER • VICTORIA • CALGARY

Copyright © 2009 Rocky Mountain Books Ltd.

All rights reserved. No part of this publication may be reproduced, stored in a retrieval system, or transmitted in any form or by any means—electronic, mechanical, audio recording, or otherwise—without the written permission of the publisher or a photocopying licence from Access Copyright, Toronto, Canada.

Rocky Mountain Books
#108 – 17665 66A Avenue
Surrey, BC V3S 2A7
www.rmbooks.com

Rocky Mountain Books
PO Box 468
Custer, WA
98240-0468

Library and Archives Canada Cataloguing in Publication
Freeman, Lewis R. (Lewis Ransome), 1878-1960
  On the roof of the Rockies : the great Columbia Icefield of the Canadian Rockies / Lewis R. Freeman.

(Mountain classics collection 6)
Previous ed. published in 1925.
ISBN 978-1-897522-46-2

  1. Freeman, Lewis R. (Lewis Ransome), 1878-1960—Travel—Rocky Mountains, Canadian (B.C. and Alta.). 2. Harmon, Byron, 1876-1942—Travel—Rocky Mountains, Canadian (B.C. and Alta.). 3. Columbia Icefield (B.C. and Alta.)—Description and travel. 4. Rocky Mountains, Canadian (B.C. and Alta.)—Description and travel. I. Title. II. Series.

FC219.F74 2009   917.1104'3   C2008-907320-7

Library of Congress Control Number: 2009920177

Front cover photo by Byron Harmon and features a teepee camp on Bow Lake during the 1924 trip to the Columbia Icefield. (Whyte Museum of the Canadian Rockies V263-NA-5994)

Printed in Canada

Rocky Mountain Books acknowledges the financial support for its publishing program from the Government of Canada through the Book Publishing Industry Development Program (BPIDP), Canada Council for the Arts, and the province of British Columbia through the British Columbia Arts Council and the Book Publishing Tax Credit.

This interior of book has been produced on 100% post-consumer recycled paper, processed chlorine free and printed with vegetable-based dyes.

To

BYRON HARMON

WHO, THROUGH HIS PHOTOGRAPHS,

HAS GIVEN THE CANADIAN ROCKIES

TO THE WORLD

ACKNOWLEDGMENT

The author desires to thank the editor of the *National Geographic Magazine* for permission to reprint portions of an article which originally appeared in that magazine.

CONTENTS

Foreword ~ ix
I The Land and the Outfit ~ 1
II Bumping up the Bow ~ 12
III Behind the Scenes of Scenics ~ 24
IV Bow Lake to the Saskatchewan ~ 41
V Bucking Mud and Flood ~ 53
VI In Camp at Castleguard ~ 70
VII The Mother of Rivers ~ 85
VIII Over to the Arctic Basin ~ 99
IX Down the Sunwapta ~ 112
X Up and down the Athabasca ~ 128
XI Back through the Snows to Banff ~ 139
Endnotes ~ 159

FOREWORD

On the Roof of the Rockies owes its existence to a chance encounter between a world-renowned photographer and a prolific US travel writer. It was late in the summer of 1920. Byron Harmon, Banff's most famous photographer, had pitched camp near British Columbia's Lake of the Hanging Glacier and was waiting for the ideal light to photograph the lake. Lewis R. Freeman, preparing to set out on a boating expedition down the Columbia, joined him. As the two chatted, Harmon unveiled his grand plan to photograph every major peak and glacier in the Canadian Rockies and Selkirks. Freeman felt it "a privilege to have met an artist who works with so fine a spirit, who has set himself so high an ideal."[1]

Freeman demonstrated his own high ideals in the adventures that provided the subject matter for his extensive travel writing. At Stanford University he had earned letters in football, baseball, tennis and track. This varsity athletic prowess set the stage for a life of travelling and exploring the four corners of the earth. Much of his life was spent on flights, cruises and expeditions to exotic places, gathering material for what would become 21 books, 15 of them published in the decade between 1918 and 1928. Freeman's propensity for adventure also led to work as a war correspondent during the 1905 Russo–Japanese War and the First World War. Other commissioned work included travelling to China as a member of a 1910 delegation of Pacific Coast chambers of commerce; to the Grand Canyon as boatman and photographer for a 1923 US Geological Survey expedition; and to Australasia in 1925 as a special correspondent with the US Navy.

When Harmon renewed their acquaintance with the invitation to participate in "the crowning achievement of 20 years spent picturing the Canadian Rockies,"[2] Freeman readily agreed. The two were to spend 70 days travelling a total of 500 miles (800 km) in the territory between Lake Louise and Jasper, Alberta, by pack train, with the principal goal of filming the majestic Columbia Icefield. It is this trip that provided the material for *On the Roof of the Rockies*.

Today, in an era when young people scarcely know what it means to wind a watch or dial a phone and cannot imagine a world without plastic or digital electronic gizmos, this book provides a refreshing look at life on the trail at a time when "our only real frills were the radio and its satellites, the carrier pigeons and the typewriter."[3] There was no instant communication via satellite telephone for these men, nor any instant direction-finding by satellite transmission to a GPS unit. The travels they required in order to achieve their photographic goals relied on the experience of guides Soapy Smith, Rob Baptie and Ulysses La Casse, along with the tales—and traces of trails—left by those few who had gone before.

Little did they know that less than a decade later the Great Depression would severely curtail pleasure travel through the Rockies and that when it resumed after the Second World War, conditions would be irrevocably altered by technological innovation. Lightweight tents and sleeping bags would replace the need for bulky canvas tents, allowing adventurers to conduct their excursions without any need for an outfitted pack train. Nowadays, at a time when the majority of alpine adventurers, each carrying their own supplies in a protective nylon backpack, travel by foot along well-marked trails with bridges across most streams, and spend the night in designated though often still primitive campsites, Freeman's work is a fascinating reminder of how quickly the nature of mountain travel has changed.

Having a string of horses transport people and goods into the wilderness along a narrow trail was a challenge at the best of times. *On the Roof of the Rockies* gives rare insight into the degree to which this challenge was magnified for men who were out on the trail for professional purposes. As an amateur photographer himself and a travelling companion of the Rockies' most notable photographic artist, Freeman was uniquely positioned to convey the strenuous effort it took to produce great backcountry images using the photographic technology of the day. He devotes considerable attention to the challenges of photographing alpine panoramas, wildlife, natural-looking camp shots and action shots along the trail.

The art Freeman refers to as "scenics"—the photographing of mountain scenes with a motion picture camera—was one of the main purposes of the trip. His explanations of what scenics is, how it evolved and how it differs from other contemporary types of photography and moviemaking are particularly fascinating in these days of digital photography. He likens the moving-picture expedition to a survey expedition, stating that:

The physical problems of the moving-picture expedition going into rough and rarely traversed country are very akin to those of the surveying party. Both may have to operate in regions of flooded valleys and snowy passes, or in dense tropical jungles, or in trackless, waterless deserts; both may be able to reach their objectives only by the crossing of broken icefields, or pushing in boats down hitherto unrun rapids. And besides the transport of food and clothing, each has to carry and protect delicate instruments, the loss of or injury to which would defeat the principal end of their expeditions.[4]

On the Roof of the Rockies aptly demonstrates how the fragile supplies and artistic demands of Harmon's photography—and the added frills of carrying a radio and passenger pigeons—added to the challenge of pack-train travel. The party carried four still cameras, an early movie camera, 8,000 feet of movie film and "as much roll and cut film for the other cameras as there was any possible chance of using."[5] And while they had waterproof containers for the movie film and Freeman's roll film, Harmon had none for his copious supply of cut film.

The men travelled in constant fear of having their supplies crushed or submerged. "A motion picture camera can occasionally be coaxed into working order again following a complete and even a prolonged submergence in water or mud," Freeman explains, "but only in the event its operator is fully competent to take it apart and put it together again. If the tiniest grain of sand or the infinitesimal part of a drop of water are left after the cleaning, trouble is certain to follow."6

The standard concern for the safety of provisions during challenging river crossings was intensified by concern for the sensitive photographic supplies. It was virtually inevitable that the dreaded event should occur: eager to avoid being submerged in the muddy trail, the pack train dove into the Athabasca River. Freeman explains that every horse, at the end of the five-foot drop from above, had enough momentum to carry him, pack and head, well under the surface. When they came up they had lost all sense of direction every horse, and so kept milling round in bunches and trying to climb on each other's backs. This was the one thing needed to make the soaking through and through of the packs quite complete ... The inevitable toll of all kinds of provisions save canned goods was serious enough, but the threat of real disaster was in the fact that all of the photographic film—exposed and unexposed—was in the boxes on the back of an especially enthusiastic swimmer called "Nig."7

By some miracle, all of the film was salvaged and not a single shot lost, though unfortunately the same could not be said for the food supply.

The hazards of fording deep, fast-flowing rivers were further exaggerated when Harmon wanted to cross a spot deeper than was necessary in order to get good photographs. Freeman reveals that at one point the crossing required by artistic demands was considered so dangerous for the supplies that the packs' contents were replaced with spruce boughs in order for the desired images to be obtained. On another occasion, camp was shifted to an impractical location in order to create a more picturesque shot.

The task of the filmmaker was further complicated by the fact that "Nature cannot be coerced nor even cajoled; she can only be humoured and waited upon. It is successful humouring and waiting that makes for the successful scenic."[8] *On the Roof of the Rockies* reveals that the unique moving picture of Mount Columbia from the north, for example, required deviating from the planned itinerary to travel a hundred miles "up and down flooded valleys and over glaciers and passes, where four or five miles a day was good progress and where there was constant risk of losing horses and packs in swimming swollen rivers."[9]

When the exhausted horses at last reached their destination, the photographers were assailed by a series of storms that obliterated the mountain from view for eight days. Finally, when supplies of food for both men and beasts became so inadequate that the guides had already departed with the pack train, the patient photographers—who had remained in camp for a few final hours of waiting—were rewarded with a half-hour of "slanting afternoon sunshine which set off the splendid peak with such a lighting as it may not have had for years."[10]

"It lasted for just 40 minutes," Freeman explained, "—ever changing but ever beautiful,—and in that time we exposed still negatives at the rate of one a minute, besides running 400 feet of film through the movie cameras. The black rectangles of paper torn from Harmon's film packs were piled up behind his tripods like the brass shells around a hard-pumped machine gun at the end of a battle."[11]

Harmon's interest in photographing the unique trail that outfitter Jimmy Simpson had pioneered across the Saskatchewan Glacier the previous year further complicated the expedition. The men took great care packing the horses. They even went so far as to fashion a crude sledge to transport the radio—though it had to be abandoned when they realized they didn't have the people power to navigate it safely across the glacier. The crossing required a long, difficult day of travel, but "the pictorial possibilities of the traverse so smote upon the artistic spirit of Harmon as to leave him for the moment gasping like a child set down in a candy shop and told to help himself."[12]

Freeman's descriptions of Harmon's efforts "to make what he had come to call his 'snow-picture'"[13] are equally fascinating. Up to that point, "neither movies nor stills had ever been made in the Rockies of a pack train trying to travel in such snows as had already closed down upon the higher passes,"[14] and Harmon was keen to take "a picture in which there was to be nothing but snow – with a few incidental horses and men."[15] This ambition meant that the party had to forsake the lower route back to Banff in favour of taking what has become known as Old Klyne's Trail.[16] From Maligne Lake they crossed Maligne Pass to Poboktan Creek, then Jonas Shoulder and Pass to the Brazeau River. They then crossed Cataract Pass and followed Cataract Creek and the Cline River to the Kootenay Plains.

Crossing Jonas Pass in deep snow was trying for both horses and men. Freeman explains that "our worst difficulty on this part of the climb, indeed, came from breaking through the crust of an earlier snowfall ... The three feet of new snow on top of this crust was enough to carry the horses only if they moved steadily ahead. The least bit of floundering put them down into the older snow, with sharp edges of the crust gashing their legs. Once through, it frequently took many yards of painful wallowing before getting back on top of the crust again."[17] Both horses and men endured, but the photographers had to be reined in by the head guide on the far side and ordered to proceed to camp for the sake of the horses rather than continuing to attempt a good movie shot.

It was thanks to the skill and perseverance of the guides—not to mention the presence of the radio—that Harmon was able to get his "snow picture." The vacuum-tube radio, powered by a pack of large dry-cell batteries and receiving signals through a long antenna strung well above the ground, had been inspired by the pleasure a similar one had brought Freeman's party on an expedition through the Grand Canyon the previous year. Practical considerations did not figure in the decision, though ultimately they prevailed.

It was actually the carrier pigeons whose presence had been influenced by practicality. In those days, so long before satellite telephones, Freeman and Harmon felt the pigeons might serve to get an emergency message back to Banff. Though they were not needed in this way, it is interesting to read about the way messages "were typed compactly on the oiled paper provided for that purpose. After pulling loose several tail feathers in an endeavour to attach the tightly folded slips as directed by experts, we finally abandoned that plan and slipped them inside of the leg-rings worn by each bird to carry its registered number. Tied with silk threads, the tiny rolls appeared in a way to ride quite securely."[18] Two pigeons were released simultaneously, and although they headed in the general direction of Banff, they never arrived at their former home. Whoever found the pigeons mailed the messages to Banff a few days later, but opted not to part with the birds.

Later in the journey, the men had the pleasure of hearing one of their pigeon messages read over the radio. This technology occupies a much more central place in the narrative than the birds do. Freeman explains that the radio began its travels in a specially made pack box that was "brass-screwed and padded on the inside [and] ... constructed to withstand rough usage, and that is why it only changed shape instead of flying apart the first time the hulking cayuse carrying it was brought up sharp as his overly broad load jammed between two close-growing pines."[19]

The radio did not always fare so well. *On the Roof of the Rockies* is peppered with accounts of the care required to transport the instrument and the perils it suffered along the way. Early on, a horse managed to destroy the box and scatter its contents widely. The men collected the pieces and returned the "battered batch of junk"[20] to the pack – only to have them completely submerged during a river crossing. Though they considered the radio a complete loss, several days later Freeman took the opportunity to "disentangle, clean and dry the pulpy bundle of wires, batteries, boxes and packing that was once a radio outfit."[21] The pattern of the radio suffering damage considered fatal, followed by the men's successful attempts to resuscitate it, continues throughout the book, highlighting the men's perseverance and passion for experimentation.

This perseverance paid off when a West Coast weather forecast of an imminent storm warned the men that their traverse of the high passes was about to be pre-empted by the weather. They resumed their travels immediately, Harmon got his "snow picture" and all made it safely back to Lake Louise.

Freeman's tales of their adventures as they photographed horses crossing fast-flowing rivers, glaciers and snow-covered passes, fiddled with the ever-engaging radio, and dealt with all the joys and challenges of late-season backcountry travel offer a delightful glimpse into the adventures of two unconventional men of ambition at a time when the world—and the world of the Rocky Mountain wilderness in particular—was on the cusp of great change. So saddle up and prepare to join Lewis Freeman and Byron Harmon on "probably the roughest continuous pack-train journey made in the Rockies since the time of the pioneers."[22]

Emerson Sanford and Janice Sanford Beck

Chapter 1
THE LAND AND THE OUTFIT

It is a strange fact that one of the first regions of the Canadian Rockies to be visited by a white explorer was also one of the last to be scientifically explored and comprehensively photographed. This is the land of tall peaks and glacier-choked valleys about the head of the Athabasca, where is found what may fairly be rated as the most striking alpine scenery of the western hemisphere.

David Thompson, the eminent astronomer explorer of the Northwest Company, who was later to run the Astors so close a race for the establishment of the first fur-trading post on the Pacific coast of what is now Oregon, crossed the Continental Divide during the first decade of the nineteenth century by following the Whirlpool branch of the Athabasca to its head and descending to the Big Bend of the Columbia by a stream he named the Portage River and which is now called the Wood. The trail thus blazed by Thompson later became the main transcontinental route of the Hudson's Bay traders between the great Canadian plains and their posts on the upper and lower Columbia. The waters of "The Achilles of Rivers," swift and turbulent but still navigable for well-handled bateaux, were available for the voyageurs to run or to breast for the thousand miles or more below Boat Encampment at the northern apex of the Big Bend.

Thus it chanced that a thin but fairly steady stream of travel flowed back and forth across the Continental Divide at the head of the Athabasca during the half century or more that the remainder of the extensive area covered by the Canadian Rockies was rarely visited by white men. It was doubtless for this reason that the narrow zone of mountains immediately under the eye of the traveller was elevated to a prominence somewhat beyond its due. Explorers and pioneers of all time have been wont to stress the wonders of what they have discovered or seen and to discount those of the regions beyond their ken.

Conservative geographer that he usually was, David Thompson estimated the height of the two peaks flanking Athabasca Pass to north and south as 18,000 feet. This was arrived at by figuring the summit of the pass as 11,000 feet according to the boiling point of water, and the tops of the peaks as 7,000 feet above the pass. David Douglas, the Scotch botanist who in 1827 named the two peaks in question Mount Brown and Mount Hooker, respectively, perpetuated Thompson's error as to height, and furthermore characterized them as the two highest peaks yet known on the continent of North America. Travellers by the old Hudson's Bay trail, unduly impressed by the twin sentinels of Athabasca Pass after their long traverses of the level plains, kept alive the early overestimates of height. School geographies studied by many people now living listed Mount Brown and Mount Hooker as the highest peaks in North America. As a matter of fact, of course, neither peak is greatly in excess of the 11,000 feet of Thompson's original estimate of the altitude of Athabasca Pass. There must be several hundred higher peaks on the continent, and two or three with summits nearly 10,000 feet in excess of the altitudes of Brown and Hooker.

Intent almost solely on finding the lowest passes and the easiest routes of travel the early traders and trappers must have been almost if not quite ignorant of the existence, but a few miles to the south, of what has since been proven to be the most extensive icefield on the continent outside of the arctic and subarctic regions of the far north. They were under the very drip of the eaves of this great continental ice shed when they tried to establish a new route to the Columbia by the Chaba branch of the Athabasca and the silver gorge of beautiful Fortress Lake which sits on the divide and drains both ways. When they failed here to find a better pass than that by the historic "Punchbowl," at the head of the Whirlpool, where the Hudson's

Bay factors were wont to rendezvous and transact the business of the East and the West in lurid week-long carousals, the glittering ice caps gleaming green against the skyline to the south must have told them that there was no use venturing farther toward the heads of the two main forks of the Athabasca. With lines of lofty peaks whose heads were rarely clear of crowns of clinging clouds, and with every mountain valley pouring down its frigid finger of glacier from a mighty mother icefield far above, they did not need to explore further to know that here was a barrier more formidable than any they had encountered to the northward. They were well advised to seek no farther in this direction. With the complete topographical data of the present day no better route could be found for the traveller by foot and by boat between Hudson's Bay and the Pacific than that blazed by the early trappers and traders. And, moreover, not to this day has a practicable way ever been found from the western drainage of the Columbia Icefield to the Columbia River, rumbling in its cliff-walled gorge six to seven thousand feet below.

The later and more extended explorations of the latter half of the nineteenth century avoided the Columbia Icefield region for the same reason as had those of the trappers and traders—passes were sought for, not barriers. Railways needed lower passes and easier approaches than had footmen and pack trains, while a cliffy gorge like that of the upper Columbia where it loops round the Big Bend was an obstacle rather than a help. For these reasons the surveyors of the transcontinental lines found that the most favourable natural routes lay far away from the lofty ice-capped plateau where three of the greatest rivers of the north took their rise. So the Canadian Pacific blasted its way through the Rockies and Selkirks from 80 to 100 miles south of the Columbia Icefield, while the Grand Trunk—later the Canadian National—found a lower and easier route as far to the north. The careful studies for neither line added little if any knowledge of the still untrodden heights of the cold ice-locked *terra incognita* between.

On account of the difficulties and dangers attendant upon

travelling with pack train in regions scarred by torrential mountain streams and exposed to the threat of slides of snow and rock, with swamps and stretches of treacherous glacial mud in the valleys, the first trails opened up in the Canadian Rockies for the benefit of the hunter and the camper were confined to the safer and more readily attainable regions near the railways. Mountain climbers, seeking for a vantage from which to attack the unscaled heights of the Continental Divide, were forced to cut their own trails through the standing and fallen timber and to scramble over rocks and slides as best they could above the line of vegetation. Much or all of such a trail might never be used again. Keen and trained observers for the most part, these courageous and indefatigable alpinists added more to the knowledge of the high Canadian Rockies during the first two decades of the present century than accrued from organized scientific exploration. Their work was fragmentary and uncoordinated, however, and it was not until the late Interprovincial Survey completed its labours along the Continental Divide between Alberta and British Columbia that definite and dependable data of this hitherto almost unknown region finally became available.

Among many valuable geographical and topographical facts revealed by the work of the Interprovincial Survey, perhaps nothing was received, even by those who knew the Canadian Rockies, with so much surprise and interest as the statement that the Columbia Icefield, formerly not known by name to one in a hundred thousand, had an area in excess of 150 square miles. Perhaps of even greater appeal to the imagination was the revelation of the hitherto little appreciated fact that drainage from this single icefield flowed to three major oceans—that it was almost certainly the only instance in the world where such great dispersion of water from a common source occurred.

The Columbia Icefield may be roughly likened to a stockily built octopus, with the main *mer de glace* forming the body and the creeping, down-crawling glaciers the tentacles. Completely surrounded by peaks varying in height from 10,000 to over 12,000 feet, the icefield itself is comparatively smooth and level, many square miles of its centre, indeed, being not more rolling than an undulating plain. The average elevation, exclusive of the tentacles of glacier which extend down not far from the 6,000-foot contour, is about 8,500 feet above sea level. Its greatest elevation is a hummock of 8,884 feet somewhat north of its centre. Here, at a point not clearly defined to the eye but probably of very small area, occurs the remarkable three-way split of the continental drainage. Where the tip inclines westerly, the water runs by the Bush to the Columbia and thence to the Pacific. The meltage from the northerly slope may run to either of the main branches of the Athabasca, and so on to the Great Slave Lake and down the Mackenzie to the Arctic. The east and south slopes drain to separate branches of the Saskatchewan which, uniting 25 miles below, ultimately mingle with the brine of the Atlantic in Hudson's Bay.

Striking scenically, unique topographically, and barely explored, the Columbia Icefield has few rivals in the world today in its attractions not only for the alpinist but the lover of the out-of-doors as well. The present article is a plain workaday account of the first attempt to make a comprehensive collection of photographs—moving and still—of this wild and wonderful region. I shall hardly need to add that all of the best of the story will have to be told by the pictures.

Even from the days when the carrying and operation of the cumbersome wet-plate outfits presented almost prohibitive problems of transport, the camera has played a prominent and increasingly important part in passing on to the older world the record of a new found land. The explorer photographs as he goes, but, handicapped by haste and the limitations of transport in a wilderness, almost always hurriedly and taking things as they come. The surveyor, when his time comes, works more carefully but only in the narrowly restricted field bounded by topography and perhaps geology. The last to come but the longest to stay is the artist—the nature photographer. He, like the settler who follows Kipling's "Explorer," "remains to occupy."

Unlike the explorer or the surveyor, the nature photographer cannot block out a region on the map and say, "When I have covered this area my work is complete." He can cover all of Nature's subjects but never reach the end of her moods. And the recording of moods—the *savagery* of the mountain torrent which grinds down and engulfs the tongue of forest that blocks its way; the *perversity* of the peak that hides its head in a veil of cloud—is his most subtle vehicle of expression.

A lifespan is all too short for the artist who would picture, either with brush or camera, a land or a race. Curtis has given the best of many decades to photographing the passing Indian of the Northwest, as has Carl Moon to his record of those of the Southwest. The magnificent photographic record of Yellowstone Park, started 50 years ago by F. Jay Haynes, is being added to today by his son Jack. Byron Harmon's successful expedition to photograph the Columbia Icefield last summer and fall is the crowning achievement of 20 years spent picturing the Canadian Rockies, but it does not mark the end of the work.

Long familiar with Byron Harmon's fine studies of the Canadian Rockies, my first personal meeting with him was late in the summer of 1920. This was at a camp on the iceberg-battered shores of that incomparable mountain gem, the Lake of the Hanging Glacier, where I had journeyed by pack train preliminary to pushing off on a boating voyage which was to carry me practically the whole length of the Columbia. In my log of that voyage I find this meeting recorded in the following entry:

"Descending to the timberline meadow where the horses had been left, we found Byron Harmon had brought up his outfit and pitched camp midway of an enchanting vista framed in green-black pines and golden larch, with a wonderful background for 'camp shots' both up and down the valley. There he was going to make his base, he said, until he found just the light that was needed to set off the Lake of the Hanging Glacier. Then he hoped to get at least one or two negatives that would do something approaching justice to so inspiring a subject. And there, working and waiting patiently through an almost unbroken succession of storms that raged in the high Selkirks for many days, he held on until he got what he wanted. It was in that quiet, patient, persistent way that he had been photographing the mountains of the Canadian West for many years, and it will be just in that way that he will continue until he shall have attained somewhere near to the high goal he has set for his life work—a complete photographic record of the Rockies and Selkirks. It is a privilege to have met an artist who works with so fine a spirit, who has set himself so high an ideal."

Something of his work—what he had done and what he still hoped to do—Harmon told me in the 48 hours we were snowbound together in the first storm of the early closing winter. That was his 16th year in the Canadian mountains, he said, and in this time he made winter and summer photographs of most of the outstanding peaks, glaciers and valleys of the Selkirks and Rockies. In four years more he hoped to have photographed them all, and then, in the light of what he had learned, he would start in and do them all over again—try to do them better.

It was on this occasion that I first heard of the Columbia Icefield. Harmon had never seen it himself, but had heard enough of it from mountain climbers who had made ascents within sight of it to rate the region as probably the finest scenically in all the Rockies. Because it was remote and difficult to reach, he was saving it for the last—the summer of his 20th year of photographing in the Rockies. With a properly equipped expedition, and by taking plenty of time, he hoped to be able to cover all of region along the Continental Divide where he would not have hitherto worked with his cameras.

Early in the spring of 1924 Harmon wrote to tell me that he had managed to maintain his photographic schedule during the preceding three years, and that there now remained only the Columbia Icefield region to picture to complete his original program for the Rockies. This he hoped to do during the coming summer and fall. Preliminary organization of the expedition was already under way, and he was writing to ask me to come along and help make the moving picture film. With a jaunt of my own already planned, which contemplated driving a small motorboat from Chicago to New York by way of the

Great Lakes and the St. Lawrence, some rearrangement of schedules was necessary to make both trips possible. Ultimately, by hard and steady pushing through the stormiest early summer the Great Lakes have known in many years, I was able to dock in New York in time to swing back to Canada and arrive at Banff by the 15th of August. The pack train was already assembled at a camp on the Bow near Lake Louise, and there we joined it the following day.

Roughly speaking, the route laid out in advance called for following the Bow River to the lake and glacier of the same name at the Continental Divide, thence paralleling the Divide as closely as topography would permit until the Columbia Icefield was reached at the head of Castleguard Valley. After a month of work on or near the icefield, including the crossing of one spur of it with the pack train and the circling round to the head of the Athabasca, under Mount Columbia, on its northern side, winter clothes and supplies would be picked up at Jasper and the return journey of 200 miles to Banff made by the best available route through the early snows. It was expected that from 10 to 12 weeks would be necessary to complete this itinerary. Enforced waits for favourable picture weather and the fact that nothing worthy of the name of trail would be available for more than inconsiderable sections was responsible for allowing what may seem like an undue amount of time to cover a total distance of not much over 500 miles.

Two features of the proposed itinerary as explained before departure impressed me—so far as my own experience of pack-train travel went, that is—as verging closely upon the impossible. These were the plan to take the horses across the icefield and the expectation of travelling for two or three weeks of the return journey through a region of high elevation, where not only the passes, but many of the valleys as well, would be deep in snow. As to the icefield traverse, I was told that such a crossing had been safely made the previous summer by the pack train of a mountain-climbing party and that there was therefore no reason to believe it could not be done again, especially as one of our packers had been with the pioneer outfit. As for the long snow journey, Harmon thought the rigours of it worth facing for the fact that neither moving pictures nor stills had been previously made of such a winter passage over the high Rockies. It would probably be attended with some losses, and it was quite possible we would have to turn back or follow one of the easterly valleys out to the plains; but, with all of the horses habituated to pawing for grass on the winter ranges, the chance was worth taking.

With every reason to expect that the trip both going and returning—the former on account of high water and ice, and the latter on account of heavy snows—would prove one of the most severe ever attempted by a pack train in the Canadian Rockies, great care had been taken that the outfit, both as to personnel, stock and equipment, should be the very best that could be assembled. With years of personal acquaintance among the guides and packers of the Rockies, Harmon had picked the three men best suited to the special needs of the expedition. These were informed in advance of the plans and itinerary, and warned that the picture work would inevitably be responsible for delays, difficulties and discomforts not usually encountered on the regular hunting and camping trips. They were all mountain men of long experience, and one of them, La Casse, was well acquainted with the region to be traversed as far as the Columbia Icefield and on to the Athabasca Glacier and Wilcox Pass.

The horses, picked long in advance, had been kept off the trail all summer to conserve their strength for the arduous and punishing work ahead. This was a most fortunate circumstance, for the long weeks of semistarvation in the snows of the early winter, with endless hours of desperate floundering in the deep snows of the high passes, demanded all their stored-up stamina to bring them through.

The photographic outfit was complete in every respect. Besides the moving picture camera, with its varied assortment of lenses, Harmon and I had each two still cameras. There was 8,000 feet of moving picture film, with as much roll and cut film for the other cameras as there was any possible chance of using. The moving picture film and my own roll films were carried in sealed tins. Harmon's huge reserve of cut films, by some oversight, was sent in ordinary wrappings, a fact which was responsible for much anxiety in the days of mud and high water.

The only real weakness in the outfit was its lack of watertight boxes and bags. The pack boxes in which the moving picture camera and its accessories were carried furnished adequate protection from the blows of rocks and trees but would not exclude water. Neither would any of the grub-boxes nor the sackings of the food supplies. As a consequence, when we were shortly confronted by unexpectedly severe high water, we lost more supplies in two days than the recent Geological Survey expedition lost in its three months' voyage through the rapids of the Grand Canyon. Even at that, however, we were never seriously handicapped by a shortage either of food or photographic supplies.

With luxuries cut to the bone as a consequence of the fact that no replenishment of supplies would be possible inside of six or eight weeks, our only real frills were the radio and its satellites, the carrier pigeons and the typewriter. The little portable Radiola was my own idea, born of the memory of the real entertainment the Grand Canyon party had had from a similar outfit a year previously. After getting the best technical advice available on the possibilities and limitations of radio under the conditions we could expect to encounter, I had bought the set in New York and brought it on to Banff. Here we had a pack box hastily built around the little black case and the block of batteries connected up for us by the local electrician.

The carrier pigeons were Harmon's idea. He had been breeding homers at Banff for a year or two but had never had a chance to try them out in the mountains. The present opportunity was too good to be missed. Besides, there was always the chance that we would be glad of means of getting out word by air in the event communication along the surface of the earth became impracticable. The birds would also prove a useful medium for reporting on radio conditions, several powerful stations having promised to make special efforts to reach us in the event the Rockies area did not prove entirely "dead," as there had been some reason to believe.

Neither Harmon nor myself knew any more about carrier pigeon technique than we did of that of radio. About all we could learn locally was from an old children's bird book, which informed that the door of the cote should be of valve-like construction, permitting the ingress of the returning bird but not the egress. Also that the boxes in which the pigeons were carried should be provided with holes large enough to admit plenty of air but not so large as to allow the birds to escape before their time. Both of these admonitions seemed quite reasonable. Less convincing was the instruction that the message should be written on waterproof oiled paper and wrapped securely to one of the bird's tail feathers. As the only line I had on this particular point was the somewhat hazy memory of the picture on a childhood valentine of a dove with a very plump packet dangling from its neck, I was somewhat diffident about disputing the practicability of the tail-feather plan. A self-constituted local pigeon expert told us a number of other esoteric practices to follow, but as the most probable of these had to do with taking along a fragment of the home roost for the bird to smell before being launched on its return flight, we did not give them serious consideration.

We did lay in a stock of oiled paper, however, on the theory that it might better resist the disintegration of possible fogs or rains. Fortunately we tried it out before starting, and so learned that even the softest pencil point would skid along on the polished surface without leaving the vestige of a mark. And that was why my little folding Corona was requisitioned to fill the breach at the last moment. A sharply struck key left an impression on the tough, smooth paper which resisted blurring even under the hard rubbing of a moist fingertip.

Chapter ii
BUMPING UP THE BOW

Driving over from Banff to Lake Louise railway station toward noonday of August 16, Harmon and I found our camp in the willows by the Bow occupied only a dozen hungry pack horses tethered among piles of hastily dumped gear. Evidently a cog of a wheel of our well-oiled plans had slipped. What the trouble was transpired an hour later when "Soapy" Smith, owner of the outfit and head packer, rode in to announce, in language more picturesque than polite, that two of his horses were missing and that he feared they were backtracking it to their native ranges in the eastern foothills. Rob the wrangler, and "Ulus" the cook were trying to trail the fugitives, but unless they came back within the hour a start that day was out of the question. Harmon, cheerily philosophical, replied that it would suit him just as well to get under way in the morning and suggested a drive up the mountain to Lake Louise to fill in the interval.

An irate packer telling the world and a considerable portion of the adjacent solar system just what he thinks of the ancestry of his strayed cayuses is not exactly at his best, from a polite and refined standpoint, that is. And yet my first impression of our head packer was unmixedly favourable. Spectacled and with the long drooping moustaches of a moving-picture sheriff, one of his friends had described old "Soapy" to me as a cross between Theodore Roosevelt and a bull walrus. It was my instinctive feeling that the man who was to guide our material destinies for the next three months combined many of the best elements of both of these virile prototypes that inclined me instantly in his favour. Too, I liked the technique of his profanity—words winged with fire but flowing with the easy, effortless inevitability of the spinning of the turbine of an ocean liner. Free natural swearing meant a well-driven, well-treated pack train. One of Nature's own swearers is also one of Nature's own gentlemen. That truth had been driven home to me through years of experience. I have never known a packer who swore freely and naturally to beat a horse cruelly. Yes, I took to old "Soapy" Smith at once, even though his name had been borrowed bodily from a notorious gambler and confidence man who had won what was pretty nearly my last dollar on the "pea-and-walnut" trick in Skagway the week before he was shot by a Klondiker from whom he had lifted a cool ten thousand.

It was nearly dark before Rob and "Ulus," tired but undispirited after their bootless search, straggled back into camp. This again proved good stuff. A man who can return with a smile after 12 hours spent in trying to pick up the tracks of strayed horses is only a little lower than the angels.

Rob Baptie was a typical Canadian Rockies wrangler. Lithe, slender, quick on his feet and with a fine bridle hand, he already had made a name for himself in the local roundups. The horse was his alpha and omega, the mainspring of his existence, the shimmer on the wings of his dreams. He could ride right down the line of the string of a passing trail outfit, call every animal by name, and tell you just what caused the rope burn on the off hind fetlock of the little flea-bitten roan straggling in the rear. He could no more keep his eyes from roving over a fine bit of horseflesh than a Broadway "Johnny" can prevent his optics from pricking a dotted line to the ankles of a passing chorus girl. All of which went to prove that Rob, like "Soapy," was of the kind that could get a lot out of a pack train without mistreating it.

Ulysses La Casse, commonly called "The Frog" because he was of French-Canadian parentage, was the all-round man of the party. Popularly credited with being the best camp cook in the Canadian Rockies, and also quite competent as a packer and wrangler as well as a guide, hunter and climber, "Ulus" was fitted to "pinch-hit" in emergency in any department of the game. He was also the only man in the party with previous first-hand knowledge of the Columbia Icefield, having been to and across the eastern side of it the year before with the Thorington mountain-climbing expedition.

With time hanging on our hands that evening after an early supper at the little hotel near the station the occasion seemed opportune for a preliminary tryout of the radio. In spite of the discouraging pronouncements of experts in Chicago, New York and Montreal, who had declared that the region of the Canadian Rockies was almost certainly a "dead" area, I had high hopes of what we were going to do with that shiny little black box o' tricks. Nor was I given pause by the fact that neither myself nor anyone else in the party knew much more about radio than we did of Einstein's Theory of Relativity. What was the use of knowing anything about it? Had not our party of comparative novices of the Grand Canyon Survey expedition given the first definite answer to the popularly and even technically accepted theory that radio could not be received in a deep, sheer-walled gorge? And now we would confound the experts again by demonstrating that the eastern wall of the Rockies did not present an insuperable barrier to the passage of the winged ethereal wave. To be sure my relations with the radio set carried through the Grand Canyon had been confined to stowing the stout yellow box containing it in the forward compartment of my boat and doing my best to keep it from being dumped out into a rapid. But I had watched to see how the ether was coaxed into giving up its secrets and felt quite confident that I could become a perfectly good coaxer myself now that I had a set of my own.

All of which showed a fine, resolute Crusader's spirit—and not very much of anything else. The ears at our headphones might have been cocked into the abysmal void for all the response my spirited jiggering of dials conjured from the unsympathetic ether. Some of the reasons for this became apparent in the morning when we took down the aerial and restowed the set in its pack box. Then even my unpractised eye discovered something over half a dozen little things that were wrong, with possibly an equal number of similar imposers of silence overlooked. A "lead" run direct to "ground" after being connected to the aerial by a hitch similar to the one Baptie would have used in securing a roped steer was one of the least of the difficulties. And that illustrates the kind of radio experts we were to begin with.

As there was too much risk of losing more animals by turning the pack horses out to forage, the whole bunch was tied up for the night with the only bale of hay procurable distributed in pitifully inadequate handfuls among the fourteen of them. They were very hungry and gaunt in the morning, for which reason "Soapy" hastened our departure in order to get them out to a good grass camp as early in the day as possible. The missing animals were replaced by two badly trail-worn cayuses from a pack outfit belonging to Bill Potts, "Soapy's" partner, which had just come in from the Yoho. Saddle-galled and weary, the recruits were far below the standard of the picked and carefully conditioned stock making up the rest of the outfit.

The addition of the radio and carrier pigeons, with their accessories, which "Soapy" had not been expecting, together with a couple of extra cases of photographic supplies Harmon had added at the last minute, swelled the loads to a volume which really demanded the increase of our pack train by two or three head. With no further animals of the sturdy stock necessary for the rough work in prospect available at Lake Louise, "Soapy" philosophically decided to divide the extra baggage among the horses on hand and let Nature take her own course in reducing it to proper proportions. I, innocently, supposed that the cryptic remark had reference to the inroads our ravenous appetites would make upon the grub supply. As the sequel proved, however, our trail-wise old mountaineer had more in mind what happens to overly bulky packs in traversing a land of muskeg and fallen timber.

Of all known methods of transport that by pack horse or pack mule is beyond comparison the roughest and most destructive.

Man-back, dog-sled, elephant, camel, boat, auto truck or aerial tramway are as nothing in their power to do harm. The crushing of lash ropes and the jolting even along an open trail are punishing beyond description. When, in addition, the pack train is headed into uncut timber, deadfalls, bogs and boulders, the destructive elements are multiplied many times over.

Into close-growing timber as soon as we turned north from the railway line, the first nine miles up the swampy flats of the Bow River were a fitting initiation for the stern work ahead. One of the first packs to be knocked under a horse's heels by colliding with the limb of a half fallen tree originally consisted of cases of jam and baking powder, with the insulated wire for the radio aerial riding between. The tangled antennæ materials were the only things to preserve their identity so as to be at all recognizable after the terrible mauling under new-shod hoofs, but the preserve-smeared and powder-dusted loops were still in a condition to lend point to old "Soapy's" wholly atrocious attempt at a joke.

"If that geesly radio don't 'jam' again when she's set up," he drawled, "this anointing ought to qualify her for broadcasting some right snappy baking receeps."

Outbreaks like that, however much deserving of condemnation in civilization, serve a distinct and important purpose during pack trouble in the wilderness by offering an outlet for pent up internal wrath which might otherwise result in violence. One begins by condoning, then encouraging, and is lucky if he does not end by embracing the habit himself.

Between mud and devious windings among deadfalls, our progress that first day was painfully slow. Now a horse was bogged to its belly; now a pack jammed tight between two close-growing trees while its bearer struggled on through; now, with all signs of a trail gone, the whole train would scatter among the timber. We were six hours making nine miles and with everyone so busy all the way finding strayed horses and scattered packs that the ever-heightening western wall of the Rockies, with its shimmering fingers of glacial ice clawing for precarious holds above the valley of the Bow, unfolded its brilliant panorama almost unnoticed.

Making camp in a drizzling rain, we took stock of the first day's attrition. With stoutly boxed and sacked stuff salvage had been almost complete. Most of the losses were suffered by things like the baking powder and dehydrated vegetables, especially where the packs had gone to pieces in mud or water. Wet baking powder was, of course, gone forever, but in the case of dampened dried fruit and vegetables the result had been to increase volume far out of proportion to the actual loss. Nature had taken her course, to be sure, but, far from acting the way "Soapy" anticipated, had probably increased the volume if not the weight of his food supplies by five or ten per cent. This meant that the horses would be packed heavier than ever for a few more days. But "Soapy's" hour was yet to come.

Finding it no longer possible to continue up the half-flooded flats of the Bow the following morning, the horses were dragged and shoved for 500 feet up the steep eastern wall of the valley to where a narrow trail had been blazed many years before. It was a desperately hard scramble for overpacked animals and far from soft work for men. Every few feet hair-poised boulders, left by the slide whose wake we followed, had to be rolled aside to give footing for the scrambling horses. Turning a rock over and propping it up to prevent its rolling down the 50-degree slope onto a pack train strung out below is an operation that requires both care and judgment, to say nothing of strength. By keeping very much in open order and out of a direct line below where active road work was in progress, we managed to stay clear of the paths of the hunks of rock which went adrift and headed a little avalanche of their own to the valley. We had all too much of that same sort of mountainside work in loose rock with the pack train before the trip was over, but, speaking personally, I was never able to arouse any enthusiasm for it. With each one of the four floundering feet of 16 horses (not to mention those of five men and two dogs) a potential starter of a moving mountainside, the feeling engendered is far from that one of comfortable placidity that comes with the reassuring clasp of the rope on the rim of a hundred-foot crevasse.

The trail which we had laboured so hard to gain proved to be the almost obliterated remains of what had been only a wretched track at the best. It was blocked in many places by fallen timber, which had to be cut away whenever presenting too high a barrier for the horses to scramble across. Congratulating ourselves on the fact that the well-drained mountainside would at least give better footing than the bottomless muskeg, we started worrying the train along through the prostrate tree trunks as best we could.

We had made about a mile when an innocuous-looking patch of moisture where a limpid springlet dribbled out of the mountainside proved that our anticipations as to the continued solidity of footing were somewhat premature. La Casse, on foot, was out in advance of the train, leading by his halter a wiry little buckskin called "The Rat." Because of his sure-footedness, "The Rat" had been given the honourable and highly responsible task of carrying the pigeon box. The latter, a frail packing case of corrugated pasteboard, was lashed so as to ride high up on top of what was already a bulky pack of bedrolls. We figured it was worth the trouble it was giving to manœuvre clear of overhanging limbs to have the box in as lofty a vantage as possible in case "The Rat" had to swim at a ford or became deeply bogged.

Now this was soundly enough reasoned out, but held good only so long as "The Rat" remained right side up. When that shifty ex-Indian cayuse found the trail under him suddenly resolving into a bottomless patch of soft mud and tried to get out of it by rolling, the unfortunate pigeons were placed in just about the position of the Hindu fanatic who casts himself under the wheels of the Car of Juggernaut. Nothing less than the quick-witted cook's catlike leap and flying tackle saved the flimsy box and contents from being rolled to a pancake under the weight of the floundering "Rat" and his heavy pack.

The birds were brought free with hardly more than an upsetting of their water can, but the case of their late bearer was more serious. "The Rat" had continued wallowing until his head was folded back under his body like that of an unhatched chicken. Then, with all four legs sticking up straight in the air, he gave up the fight and began resignedly to strangle. Rob, the wrangler, had already thrown a hitch over the forelegs and started pulling down the trail when "Soapy," dashing onto the scene from above, lassoed the hind legs and set his mount pulling in the opposite direction. It was decidedly rough on "The Rat" in a sense, this Spanish Inquisition treatment, but in the end it proved his salvation. Pulling himself together before he was completely pulled apart, he kicked free from "Soapy's" hitch and, turning a complete somersault, landed on his feet and came bounding along in the slithering wake of Rob.

"The Rat" was a tough little brute, and where an ordinary horse could hardly have been less than drawn and quartered by so terrific an experience, he showed almost no effects of it for a while. Later it became evident that there must have been some injury to his back, for he was never able to raise his rear hoofs more than a foot from the ground when trying to kick. Yet even this trouble, whatever it was, never affected his usefulness as a pack horse. He continued to bear the heaviest of loads even through the deep snows of the last weeks of the trip.

Tall, slender extremely dense timber screened all but evanescent glimpses of a grey-green sheet of water widening across the valley floor below. This was Hector Lake, named for the physician and surgeon with the Palliser expedition, which made the first scientific exploration of the passes of the Canadian Rockies in 1858. It is fed by a stream from the Hector Glacier, which extends down from the summit of the Waputik Mountains, as the main chain of the Rockies is called in this region. It drains to the Bow, which, at high waters, covers the low level valley flats to the south and east with a lake of its own.

Passing out of the heavy timber, a half mile among the chaos of piled rocks and upended tree trunks where the snowslides of a thousand centuries had left their accumulated scourings at the foot of the mountain, we crossed the boulder-choked channel of Mosquito

Creek and descended again to the flats of the Bow. In spite of several fordings of the deep, swift river, the going here was the best we had enjoyed since leaving Lake Louise. This was due principally to the fact that, with the valley floor sloping at a sharper declivity, more water had been drained out of the clinging blue-grey glacial silt with which it was paved.

First and last, glacial silt was the most annoying scourge we encountered on the whole journey. This main by-product of the grinding of the mills of the ice makes the muddiest mud, as well as the dustiest dust, with which I have ever had personal contact. Like the Mills of the Gods in the poem, those of the glaciers, though they grind but slowly "yet they grind exceedingly small." The dust is so impalpable that it will filter through the closest woven canvas; the mud, at its worst, offers no resistance whatever to the downward passage of the foot or the body of horse or man. Once into it, one goes right on to bedrock, unless he has help, or is so fortunate as to encounter some extraneous body like the trunk of a tree.

Our packers claimed that glacial silt had one use. It was a wonderful abrasive—had no equal, in fact, for putting on strops for sharpening razors. Not having shaved during the trip, I had no chance to test the pestilential paste on razors. As to its abrasive effect on tempers, however, I can testify from a full heart.

Following the winding ribbon of the Bow until mountain meadows, gay with flowers, gave place to the narrow and precipitous canyon which drains the lake above, we were finally forced to ascend a series of sloping benches to the north for another miserable stage in bogs and fallen timber. Climbing slowly but steadily, we skirted the attenuated finger of the lower lake, passed the swift-flowing narrows above, and came out at the end of the afternoon upon the firm pebble-paved beach looping in the easterly arm of the upper or main Bow Lake.

The scene, especially as it burst upon us after the terribly wearing struggle with an all but exhausted pack train in the black inferno of burned timber and mudholes below, was of an unearthly loveliness.

To our left was the weird Crowfoot Glacier, clutching with icy talons the precipitous slopes of towering Bow Peak to keep from falling into the foam-white rapids of the narrows below. To our right, a long gently sloping wedge of meadow, dark green and brown and mottled with the shadows of low-lying bushes and clumps of snow-stunted pines, led up to the broad notch of Bow Pass, through which could be seen the pinnacles of the snowy peaks across the Saskatchewan floating, sun sharpened, against the blue haze beyond. Ahead, across a milky-jade lake surface gently ruffled by the evening breeze, was a wall of ice and rock culminating in the solid mass of Bow Glacier, which reared its bottle-green snout above a grey boulder-fan streaked with glittering runlets of tumbling water.

Two great waterfalls, flanking the face of the glacier to left and right, showed perpendicular shafts of gleaming white, round and solid, like the marble pillars supporting the arch of a Grecian temple. The slanting afternoon sunlight formed concentric circles of rainbows in the mist-whorls rising from the foot of the twin cataracts, before striking through to prick with opalescent gleams the dancing wavelets of the lake below.

It was a pity that the day's travel could not have been brought to an end then and there, for taking to the mud and the timber again was a sad anticlimax after the sudden unfolding of that inspiring vision. But a campsite in the northern Rockies demands not only a fairly level and comparatively dry place for the tents, but also good grazing for the horses within reasonable distance. With none of these essentials available where we were, there was nothing to do but push on until they were found. The fact that we had already been eight hours on the road, where four or five was all a pack train ought to be required to endure of such travel, had nothing to do with the situation.

The mountainside sloping to the east end of the lake proved to be honeycombed with bubbling springs, clear and beautiful to look upon but potential morasses for the weary horses. At the end of a quarter of a mile, with half of the animals down and all of them near the end of their strength, we gave up the fight with the mud and dragged them, one at a time, down to the lake. There was no beach for the next mile, but with the bottom solid and the water not over ten to twenty inches deep, it was possible to make slow but steady progress.

Splashing and floundering along over slippery rocks, the leaking mountainside was skirted, only to find the lake bottom becoming soft and soggy where the little valley ran back to the summit of Bow Pass. What had looked like a pretty meadow a mile away turned out to be brushy muskeg, with a narrow, steep-banked stream winding back and forth across it like the wake of a wounded snake.

With several of the horses ready to quit every time they bogged down, we were nearly an hour wallowing our way across the last half mile. That behind us, we pushed through a forest of fine old spruce to a protected and beautiful campsite a couple of hundred yards back from the north side of the lake, with Bow Glacier, rosy pink in the sunset glow, reared against the southwestern skyline.

Rain in the night, turning to snow for an hour or two after dawn, was followed by a day too lowering and overcast to make work in the picture line practicable. Snow on the 20th of August was a significant reminder of how near winter stalks in the northern Rockies, even in midsummer. Stringing the radio in the afternoon, with the aerial between two lofty pines, we were awarded by vigorous wails from the ether which gave promise of better things once there was time to make a careful set-up. We had learned enough to reassure us on two important points: that this section of the Rockies was not in any sense a "dead" area, and that the terrific banging it had received had not seriously impaired the usefulness of our little receiving set.

Quite our most interesting radio reception of the day, however, had nothing to do with ethereal messages. It came about through the fact that, during the forenoon, salt had been thrown for the horses on the identical patch of grass where the radio was set up later in the day. Twice or thrice in the course of the afternoon a bunch of the horses straggled down from the meadow above for another grateful lick of salinity. The last time they were accompanied by two splendid whitetail does, the first deer we had seen.

Taking our cameras and backing off unobtrusively into the timber, Harmon and I left the salt lick, and incidentally the radio, to our graceful lady visitors. The strange-looking and -smelling box seemed to have a greater appeal to the curiosity of the pretty pair than did the salt to their palates. Once fully aware of it, they stopped licking for a while and confined their activities to stepping round and round the wonder in narrowing circles.

By the time we had slipped and slid along into a position suitable for pictures, one of them had satisfied her curiosity and gone back to munching the salty grass. The other, when I snapped at a hundred feet, was standing with extended neck, her sensitive nose sniffing not many inches away from the dials. Stalking still closer, I was about to snap again when a snort, ripped out at the edge of the timber, sent both of our afternoon-salt guests bounding away. Their lord and master, evidently becoming suspicious over developments, had ordered his ladies home.

Notwithstanding the character of the offering which first attracted our visitors, my photograph will attest that this story need not be taken *cum grano salis*.

Chapter III
BEHIND THE SCENES OF SCENICS

ANOTHER NIGHT WITH SNOW FLURRIES gave way to a day of ideal mountain photographic weathers—sunshine and squalls with their butterfly chases of lights and shadows. As this marked the beginning of our work with the cameras, a few words about the making of scenics may be in order before going on with the story of the expedition.

With the veriest tyro of a movie fan sapient of all the tricks and intricacies of photo-play production, from the simplest of double exposures to the mechanics of the opening up of the Red Sea to engulf the armies of Pharaoh, there is still very little knowledge of the "inside" work of shooting scenics.

The generally accepted idea appears to be that a scenic moving picture is made by simply going out and travelling among beautiful mountains and lakes and rivers and exposing film on the best of them as the cameraman goes along. This primitive idea of the genesis of the scenic I found held even by a distinguished engineer who had described to me in detail how Notre Dame cathedral in the "Hunchback" had been shot in a combination of solid sets and paintings on glass. The fact that an effective scenic shot might cost more time and trouble than the construction of a photo-play Babylon or Paris had never entered his head.

The reason that the public has pushed behind the scenes in the case of the photo-play without ever suspecting that there was any "behind" to the scenic is doubtless due to the fact that it knows that the former is the result of an attempt to create an illusion, while the latter is merely an attempt to hold a mirror up to Nature. This view is correct in the main, but what the general public appears never to have sufficiently understood is the fact that, in many instances, holding the mirror up to Nature may require quite as much preparation and trouble as the creation of an illusion. Nature cannot be coerced nor even cajoled; she can only be humoured and waited upon. It is successful humouring and waiting that makes for the successful scenic.

As a by no means extreme example of what waiting on Nature may involve, I might mention the work, time and privation incident to getting a moving picture of Mount Columbia some weeks later on the expedition on which we were now embarked. Knowing that this most beautiful pinnacle had never been photographed from the north even by ordinary cameras, we planned a deviation from our original itinerary that would take us to the head of the Athabasca, which drains toward the Arctic from the Columbia peak and icefield. This called for a hundred miles of travel up and down flooded valleys and over glaciers and passes, where four or five miles a day was good progress and where there was constant risk of losing horses and packs in swimming swollen rivers.

Reaching timberline at the head of the Athabasca with exhausted horses, we were assailed by a series of storms which completely shut out the mountain from view for eight consecutive days. At the end of that time, with the horses all but starved from lack of grass and our own provisions reduced almost to the vanishing point, we were rewarded by half an hour of slanting afternoon sunshine which set off the splendid peak with such a lighting as it may not have had for years. We were on reduced rations for a week as a consequence of that vigil, and some of the horses never did regain their full strength; but we got our pictures, both stills and movies.

The photographer on the "lot" has always in his favour the fact that he is working with sets built for no other purpose than to shoot most effectively under ideal lighting conditions. The scenic photographer has to take Nature as he finds her, and it is undeniable that some of the most beautiful natural manifestations have not good "screen faces." There are features like the Yosemite, Kaieteur Falls and the

Matterhorn which, in the picturesque parlance of the cameraman, "shoot like a million dollars." With a composition as perfect as that of a built-to-order studio miniature, the photographer has only to wait for the clouds and sunshine which will give the light and shadows to register moods.

There are other great natural wonders, like the Grand Canyon of the Colorado or the Yguazu Falls, which are always disappointing in the pictures, especially to those who have seen them as they are. They make interesting views, but—principally because they can be pictured only in bits—fail to put their personalities onto the screen.

The magnificent Fall of the Athabasca is a case in point from our present trip. This beautiful cataract, falling from an open valley into a deep gorge, is bathed in so unequal a light that one part of its shimmering shaft of leaping water cannot be photographed properly without greatly over- or underexposing all the rest. This condition prevails at low and middle stages of water. At flood stage the great horseshoe of tumbling white is completely obscured by the heavy mists floating up from the surges in the rocky gorge below.

Save for the occasional bursting of a dam or the breaking of the waves of a storm against a seawall, Nature's greatest action pictures are rarely caught by the camera. The eruption of a volcano is occasionally filmed, but only from a safe distance. Eruption, earthquake and cyclone pictures are usually pictured only in the aftermaths of disaster. This is also true of cloudbursts. The man on the spot seldom has a camera of any kind, especially one for taking motion pictures. Or if he has, the light or his vantage may be wrong. A still picture which I took from a very favourable position of a tremendous cloudburst pouring over a castellated pinnacle above the Grand Canyon of the Colorado came out on the film no more savage than a summer sunset.

I have never heard of a good picture of a major avalanche, though a cameraman who accompanied me to the head waters of the Columbia in the Selkirks of Canada missed by a hair the chance of a lifetime to make such a shot. Or perhaps I should say that he missed it by an echo, for that was what was really at the bottom of the heartbreaking failure. This occurred at the Lake of the Hanging Glacier, whither we had journeyed by pack train to make the preliminary shots for the filming of a boat trip from the source to the mouth of the Columbia on which I was about to embark.

The setting for a movie was incomparable. The "Hanging Glacier," a mile wide across its face, closed the farther end of the lake with a 200-foot wall of solid ice. Back of this was an ice-sheathed cliff 2,000 feet or more in height, crowned by a snowcap, white and smooth as a marble dome, sparkling against a sky of deep azure. The lake itself was a glittering emerald nestling in a setting of glaciers and ancient snows, with its glassy surface reflecting the bizarre shapes of floating icebergs and the reversed images of towering walls.

In an endeavour to get some life and action into the movie shots we were trying to set some of the floating icebergs in motion by exploding sticks of dynamite under them. It was this that was responsible for creating what I am inclined to believe was the greatest opportunity ever presented to a moving picture operator to film one of the most stupendous of Nature's manifestations.

The brink of the great ice cap must have been all but ready to fall by its own weight. The shock of the detonating dynamite provided just the sort of jar that was necessary to start it ahead of Nature's schedule. The whole far end of the lake suddenly seemed to be falling in. The tumbling cap formed an unbroken cataract of glittering ice and snow—a mile wide and half a mile high—right down to the level of the glacier. And the jar of this avalanche set the glacier itself vibrating, so that all of the lakeward wall of it cracked off and fell into the water to form a fresh phalanx of wallowing icebergs.

A strange trick of alpine acoustics was responsible for the fact that this tremendous picture was not perpetuated upon celluloid. The cameraman, with his instrument already set up to catch the bobbing of the imminent bergs as the dynamite awakened them to life, would have needed but a few seconds to swing round his long-focus lens and train it upon the great slide. Unfortunately, those precious moments were lost trying to locate the disturbance on the side of the valley from which its thunderous echo caromed down to the berg-battered beach upon which the camera was set up. By the time snow-glare-blinded eyes were squinting in the direction of the real avalanche the show was over and the opportunity of a lifetime gone. Only a few thin trickles of snow were streaking the face of the cliffs when he finally swung his powerful telephoto upon them, and even these had ceased before he had found a focus.

It was no end of a pity. I was present when some of the largest *valangas* of recent years were started in the Dolomites by the Italian artillery during the late war. Yet not all of these combined would have equalled a tithe of the weight and volume of the titanic avalanche of ice and snow set in motion by our innocent stick of dynamite at the Lake of the Hanging Glacier.

And this illustrates another salient difference between photo-play and certain kinds of scenic photography. On the "lot," or even on location, if a scene is not shot to the liking of a director, he simply roars his classic "Rotten! Do it again!" and the whole action is repeated. This can be, and often is, gone over, a dozen times if necessary.

But Nature cannot be handled in this summary fashion. Fixed and immutable things, of course, like mountain and waterfalls, stay right where they are, to be shot again and again as long as the light holds good. But slides that have slid are gone forever, and so, also, may be lost the sunset of a thousand that has to be passed up because the horse with the movie camera is down in the mud.

Nor can a sheer-walled rapid, like so many I have run for a movie in the Grand Canyon of the Colorado and the Columbia, be shot again because the camera balks during the first attempt. Nor yet is it often practicable to risk horses and packs a second time in a dangerous river ford. In all trail and river pictures of this kind there is no "comeback."

The failure of the first shot means the loss of the scene and possibly a break in the continuity of the scenic.

The first scenics were no more than a series of disconnected movie shots, screened about as one would turn the pages of a Kodak album. These were followed by the so-called "travelogue" type, with a man walking through the pictures and pointing out the various objects of interest after the fashion of an animated Baedeker. This was an improvement, but the floorwalker manners of the cicerone often made the showing singularly reminiscent of "our Mr. Cohn" taking visitors over the newly opened wing of a department store.

If there ever has been found a man who can "demonstrate" the panorama from a mountaintop without going through the wooden poses of an Indian cigar sign I have yet to see him on the screen. A world of snow and ice towering above a stream-streaked valley is on a bit too vast a scale to be shown off like a real estate subdivision or a stack of Persian rugs.

A few odds and ends of outdoor views (not inaptly called "screenics") are still occasionally shown by the old bargain counter method and to the accompaniment of the wooden "ain't that an eyeful" gesture. But the present-day maker of scenics has adopted the simpler, more intimate and less undignified plan of making the person who sees the picture screened a sort of silent partner in the expedition. This system is at its best where the scenic contemplates covering a barely explored region such as the one into which our present expedition was about to penetrate—a land which the public knows little or nothing about.

A map, with a few words of history and topography, may be screened as an introduction; then the departure, followed by connected and carefully correlated shots which tell the complete story of the expedition from start to finish. The outstanding scenic views will always be the main thing, but these will be linked by a series of shots showing the troubles, and perhaps the tragedies of the trail, the routine and humours of camp life, with something also of the problems of transport and their solving.

Five or six reels of a scenic intelligently made along these lines should give one seeing it screened a clearer idea of a region of which he may never have previously heard than many weeks poring over books covering the identical section. The work and worries, the thrills and griefs, of trying to introduce life action and a human touch into such a modern scenic will unfold with the story of our expedition to "The Mother of Rivers."

The physical problems of the moving picture expedition going into rough and rarely traversed country are very akin to those of the surveying party. Both may have to operate in regions of flooded valleys and snowy passes, or in dense tropical jungles, or in trackless, waterless deserts; both may be able to reach their objectives only by the crossing of broken icefields, or pushing in boats down hitherto un-run rapids. And besides the transport of food and clothing, each has to carry and protect delicate instruments, the loss of or injury to which would defeat the principal end of their expeditions.

In this latter respect the film man has by far the more difficult problem. All of the standard motion picture cameras are so much heavier than the transit, alidade or other instruments of the surveyor that the difficulty of protecting them is increased many fold. The best that can be done in other instances is to provide strong, padded watertight cases, and, wherever possible to have these containers transported by hand at the points of greatest menace—bad rapids if travelling by river, broken slides of rock and fallen timber, if moving by pack train.

A motion picture camera is as delicate as a watch and several hundred times as heavy. Carefully cased, it will survive all ordinary hazards of wilderness travel short of the rolling of a horse down a precipice or the smashing of a boat in a rapid. Once out of its protective boxings, however, as it is at all times when about to be used, a fall of over a foot or two to anything save soft earth or snow will almost certainly wreak injury beyond all but factory repair.

A springing of the metal frame too slight to be detected by the eye, or the least disalignment of the intricate cogs, may cause a jamming of the film that will ruin shot after shot. Extraordinary care at all times in carrying the camera is the best that can be done to save it, and it is astonishing the amount of difficult and even dangerous climbing that can be done with the heavy, cumbersome instrument without enough of a misstep to result in its permanent injury. Equally remarkable, on the other hand, is the way a movie camera will become obstreperous for no apparent reason at all save pure cussedness. Trouble of this kind almost invariably occurs at the beginning of an important shot—only too often indeed, one which there is no chance to make over again.

A motion picture camera can occasionally be coaxed into working order again following a complete and even a prolonged submergence in water or mud, but only in the event its operator is fully competent to take it apart and put it together again. If the tiniest grain of sand or the infinitesimal part of a drop of water are left after the cleaning, trouble is certain to follow. Even the lenses have to be taken apart and thoroughly dried to get rid of possible moisture which will later form a blurring fog.

Extremes of heat and cold are among the most difficult things with which the maker of scenics has to contend. In our boat voyage of 1923 through the Grand Canyon of the Colorado, for instance, temperatures running as high as 120 degrees Fahrenheit so softened the film that it was impossible at times to reload it without injuring the emulsion. On the present trip, on the other hand, there were mornings when the mercury dropped down to 20 degrees below zero.

The delicate adjustments of the lenses could not be made with mittened fingers, yet the ordeal of touching chilled metal which seared the skin like contact with a hot stove was really one of the least of Harmon's many troubles. His worst difficulty was that of preventing the fogging of his many lenses and filters. A camera brought from a fire-warmed tepee on a zero morning would always condense moisture which could only be completely removed when temperatures of the metals and glasses were reduced to that of the air. This trouble was minimized by leaving all of our cameras, carefully covered, out of doors during the night, but the difficulties which arose from the fogging of lenses from our own breaths were always with us on cold days. I fogged a filter on Castleguard (and incidentally spoiled an important picture) with the moisture from the same breath with which I was warning Harmon to keep his own frosted exhalations off his six-inch telephoto.

The presence of abnormal amounts of static electricity in the air of high mountain and desert regions is responsible for quite as much trouble to the moving picture man as to the operator of a radio. The friction of the film on metal creates brilliantly flashing sparks which are registered on the negative in a form strikingly suggestive of the miniature bolts of lightning which they really are. A specially coated anti-static film reduces the risk from this source materially, as does also the practice of minimizing friction by placing small strips of felt at all points where the celluloid strip passes over metal. In spite of all precautions, however, these bolts from the hand of a baby Jove are likely to descend at any moment to jazz their forked trail across an otherwise perfectly exposed strip of film.

The physical difficulties of transporting and protecting his outfit and of making technically perfect film overcome as far as possible, the maker of a scenic can turn his attention to the picture itself. Where the shooting of a motionless subject, such as a mountain peak, a cliff or a valley, is concerned, one might think that the thing could be done quite as satisfactorily with a still as with a movie camera. This is, indeed, quite true if the peak alone is considered. A stereopticon slide would show it off just as effectively as would a strip of running film, and without the flicker. But it is in its ability to introduce life and action into the picture that the movie has the advantage.

The "slow-cranking" of the clouds blown across the half-revealed summit of a snow-capped peak makes it possible for the movement to be greatly accelerated by running the film through the projector at the normal rate on the screen showing, producing the effect of a breaking storm so often resorted to in the photo-play shots. Waving trees, nodding reeds at the border of a wind-rippled lake, or a flowing stream add immeasurably to the effect of the snowy peak towering in the background in majestic immobility. Finally, there is the use of human figures in the foreground.

I have mentioned the way in which many scenics are marred by the wooden posturings of "demonstrators" who give the impression of trying to auction the screened landscape at so much a yard. Unless a man can stand out on a rock and look at a mountain or valley in more or less the same way he would if he knew there was no clicking movie camera within a thousand miles, he had much better be kept out of the picture entirely. "Learnin' 'em to act nateral" is one of the most baffling problems confronting the scenic director.

Admonishing a packer or a boatman just to "be natural" is giving a "counsel of perfection" no more practical to follow than Jerome K. Jerome's direction to the young man who would be a success in society. It is no less difficult to be natural to order before a cranking camera than it would be to "adopt an easy and pleasing manner, especially toward ladies," as the British humourist suggested.

If it is simply an ordinary routine trail or camp shot, such as unpacking the horses or pitching the tepee, it is usually enough to get the background and the lightings right and tell the men to "Make it snappy!" The accelerated action concentrates their minds on the work in hand and keeps it off the disconcerting glare of the one-eyed box. But taking them out on a skyline cliff, between the Devil of the camera and, the Deep Sea of Nature, and keeping them still the simple, unaffected children of the wild is quite another matter. Self-consciousness supervenes instantly on a solemn occasion of that kind, and unless this is exorcized one's lithe, graceful, picturesque woodsmen become straightaway wooden cigar-signs.

The obvious remedy, of course, is to make them forget themselves, and to this end I have found nothing more effective than a run of inconsequential chatter directed from anywhere behind the field of the lens. Almost any kind of patter that comes into the head will do as a rule, but funny stories—especially those of a subtle character— are best avoided. If the men get the point they are likely to be so tickled over it, and incidentally their own cleverness, as to laugh right out of their parts. If they miss the point, on the other hand, they are certain to become even more wooden than if Nature had been allowed to take her course.

A story which proved a boomerang and broke me for good and all of the narrative plan of offstage recitative to make the units of my human foreground forget themselves and "act nateral," was told on a peak of the Selkirks in making the preliminary Shots for my voyage down the Columbia. Ninety feet of a wonderful sunset across a purple-shadowed valley had been cranked, with but ten left to go, when my story came to an end. Instantly the look of "rapt admiration" on the face of my head packer and leading man was replaced by a broadening grin as the point drove home and began to tickle his risibilities. The grin gave way to a chuckle, and that to a thunderous guffaw and the announcement that the yarn reminded "of the dancehall gal down in Revelstoke that had one of them fluffy white pups allus wore under the arm." Of course he stepped right out of his part of "scenery-awed woodsman" right then and there to tell what "Wild Bill of the Big Bend of the Columbia" did to that "gol dern pup"—and of course the whole shot had to be made over in a fading light.

There is no end of the things that can turn up to throw a monkey wrench into the machinery of a perfectly oiled and smoothly running scenic shot. On one occasion on our present trip it was a big bull moose; on another it was a small dead mouse. The moose, lumbering down from a slide on the mountainside, stampeded a part of the pack train that was being bunched for a shot up the Athabasca River with Mount Columbia in the distance. The sad part of it was that, besides frightening the horses out of the picture, the moose—a magnificent specimen with an especially fine head—swerved off into the timber just before getting inside the focus himself.

The intervention of the mouse was even more serious, but that story I will tell in its proper sequence.

The handling of a series of scenic shots where it is desirable to introduce a touch of educational interest is well illustrated by the manner in which we filmed the panorama from the summit of Mount Castleguard, overlooking the great Columbia Icefield. This is not only one of the most beautiful mountain views in the world, but also one of the most interesting. Besides looking out on a veritable sea of lofty peaks notching the skyline to a distance of over a hundred miles in every direction, one may fix his eye upon almost the exact point in the middle of the 150 square miles of the great *mer de glace* where the drainage to three major oceans divides. As I have stated in an earlier chapter, this is the only place on the face of the earth where three great rivers, each flowing to a distinct and separate sea, head at one topographical apex. Our problem here was to give as graphic portrayal as possible upon the film of this unique geographic phenomenon. First, in order to make the most of the scenic effects, a complete 360-degree panorama was shot, using the lenses and filters best calculated to bring out the lights and shadows of the surrounding world of ice and snow. Three of the packers who had made the ascent with us were then grouped in the foreground of a shot directed toward a point where a break in the western rim of the icefield indicated the gorge where the Bush River drained to the Columbia.

In a similar way the rifts draining to the Athabasca and Saskatchewan were shot in turn, the men in each instance looking toward the focus of interest and gesturing in a way intended to indicate that they were talking and thinking of the remarkable three-way dispersion of the meltage from the great icefield. This made place for titles telling how the waters of the western drainage ultimately flowed to the Pacific, those of the northern to the Arctic and those of the eastern and southern slopes to Hudson's Bay and the Atlantic. It also gave opportunity for subsequent cut-ins of bits of scenes from the three great rivers and oceans, showing the mental pictures conjured in the minds of the men as they looked down on the rounded hump of ice and snow forming the strange and wonderful continental apex.

Shots of a humorous character occasionally go very well in camp scenes but are usually calculated to detract from the dignity of a great mountain, valley or other natural setting. We made an exception to our rule of not making shots of this kind, however, when one of the packers, twirling his rope and leaping through the loop of it on the brink of a 3,000-foot cliff, offered an opening for some such title as "The Highest Jump on Record" that was too strong to resist.

Little shots of this kind, as long as they stop short of horseplay and comedy slapstick stuff, are frequently desirable by way of relieving the tension that is likely to be strung to the breaking point by too long a footage of straight unadulterated "Ain't Nature grand?" scenes.

Camp-shots, because they can be made on days when the light is unfavourable for photographing landscape, offer few difficulties save on the score of variety. Packing, setting up tents, tossing flapjacks and similar routine work have been done to a finish; also campfire shots by flare-light. I myself, on my Columbia River trip, introduced the only drastic variation ever filmed in a campfire scene when I walked into the picture and sat down on a smouldering log instead of the roll of blankets which had been placed for me. And even the fine frenzy of that highly unconventional action was ruined for the screen when my flying leap for the river carried some very expressive and unpremeditated pantomime out of camera range.

Pies and cakes, produced as by the wave of a magic wand from an ash-covered Dutch oven, are "surefire" stuff in camp shots; also such little touches as candles improvised from twisted strips of bacon decorating a birthday cake.

Some lively action shots are always provided by a bucking pack horse, provided you are so fortunate as to have one in the outfit. Ornery animals that have to be thrown before shoeing also make for hectic action. The best potential action shots on the trail—on the occasions when the real grief occurs—are not often made. A horse with its head doubled up under its back in the mud and in imminent danger of breaking its neck, almost invariably demands the instant help of all hands including the cameraman; likewise an animal with a valuable pack about to be sucked under a logjam by a 20-mile current or down between rolling boulders at a rocky ford.

It is also practically *de rigeur* in scenic trail etiquette to throw a rope or extend an alpenstock to a man down a crevasse before starting to crank film on his predicament. As, in actual practice, it is usually the preoccupied and temperamental cameraman himself who drops into the hole in the ice, this nice discrimination is not often demanded.

Dogs are frequently very effective in trail and camp shots, but unless highly intelligent are likely to make a deal more trouble than they are worth. We had one dog on the present trip—a collie-husky crossbreed—which performed brilliantly in riding packs across swollen streams, registering the whole gamut of canine emotions with the radio headphones over his ears, and even dragging the receiving set itself part of the way across the glacier on an improvised sledge to save the delicate instrument from the almost annihilative rigours of pack train transport.

"Buster" was the motif for several hundred feet of live, snappy film, or just about the same footage that was completely ruined by his mate, an Indian-bred mongrel which had a special penchant for licking lenses and capering figure "8's" between the legs of men planted in the foreground of a scenic shot to register "rapt admiration" or "awed wonder." No man, least of all a packer with his high spirit and prima donna temperament, is capable of registering "awed wonder" while teetering to regain a dog-destroyed balance on the edge of a 1,000-foot cliff and swearing in cataclysmic Cree and English at the cause of his troubles.

Cataracts and rapids have enough action of their own for a movie, but, unless one is striving to perpetuate a mirror effect, the surface of a lake shoots better when stirred by a breeze. Inasmuch as a wind-machine is not among the "props" that can be carried by pack train or boat, one usually has to wait until Nature is in a propitious mood if it is a breeze-rippled lake that is the *desideratum* of the moment. Effective agitation, though hardly similar to that of wind action, can, however, be obtained in various ways.

One may, for instance, take a leaf out of the book of Xerxes, who gave the Aegean (or was it the Bosporus?) a good beating with whips because it had engulfed some of his transports. Lashing a lake surface with the limbs of trees—the actual lashing beyond camera range, of course—will throw up a very merry little dance of ripples to brighten a backlighting effect or set nodding a lacustrine fringe of reeds. If more violent artificial action is demanded there remains dynamite, which is easy to transport and safe enough to handle if reasonable care is exercised. Besides the accidental starting of a whole mountainside sliding, as I have already described we produced one very effective close-at-hand avalanche by the use of dynamite on my Columbia expedition. The details of this stirring episode have been told elsewhere.[1]

Wild animal photography, both still and movie, is in a class of its own, though a few shots of game in its native haunts cuts into a scenic very effectively. It demands patience and nerve on the part of the operator, both in large measure. As in hunting, success depends very largely upon luck. Indefatigable climbing is the main thing in shooting goat or sheep, either with gun or camera. In the case of the latter, however, many a heart-breaking clamber is stripped of reward by impossible light or backgrounds. Deer, elk, moose and caribou require interminable stalking, and sometimes driving, to get them in effective pictorial surroundings.

Bear are usually not hard to manœuvre into a position for a movie shot, nor, ordinarily, is there more than a negligible element of danger in it. A she-bear with cubs, however, albeit quite the finest subject conceivable, is a creature of uncertain temper and had best be kept at a distance or covered with a rifle. How we made some highly successful movies of a very pugnacious old lady bear and her two cubs by shooting them across an impassable gorge with a telephoto lens I will tell in a later chapter.

I have seen a Nepalese tiger, charging a cranking cameraman, stopped by a covering rifle; also a movie of a very similar occurrence with an African rhino. The one onslaught which I have seen that nothing could stop, and which came within a hair of blotting out the life of the photographer who stood his ground by his camera, was that of a mountain goat. That the goat had been dead for 24 hours, and was frozen stiff as well, did not make the affair a whit less serious. The incident occurred in the Selkirks, in 1920, and the cameraman was Byron Harmon, my associate of the present expedition. I was working with my own outfit at the time and so figured in the near tragedy only as a spectator.

Harmon had been trying vainly for several weeks to make a film showing the stalking and shooting of the goat among its native crags. Several fine specimens had been brought down with a rifle, but not one of them under conditions of lighting and setting that made the resultant film at all satisfactory. Finally ally he conceived the idea of making the killing part of the shot in bright sunlight with a goat already dead. To this end a big "billy," shot on the cliffs a thousand feet above the glacier at twilight, was left where he had fallen. The next day they proceeded to enact a "shooting" which could be adequately transferred to the film.

While my party was filming a scene in an ice cave at the glacier which was to be the introduction of my Columbia River picture, Harmon had finished setting the stage for his goat scenes. He planned to make two shots—one of his packers firing at the goat—propped up in a lifelike position behind a ledge of rock—and the other of the body of the goat falling to the glacier.

The "killing" of the goat went off quite satisfactorily, both in long shots and close-ups. A concealed string to the goat's hind leg insured a realistic toppling over even after two bullets pierced the whiskered head without budging the stiffly braced frozen frame. The hitch came at the shooting of the old "billy's" thousand-foot "leap of death" to the ice of the glacier.

Harmon, in setting up his camera as near as he could to point where the "leap of death" was going to culminate, had made his estimate not wisely but too well. From where I watched through my binoculars it looked as though the hurtling body was almost certainly going to strike both camera and operator. Nor did the sequel prove my judgment wrong. Harmon, suddenly alarmed by our shouts and the swift increase of size of the white ball in his finder, ducked just in time to turn a solid collision into a sharp rap from a flying hoof or horn. Some other section of goat anatomy knocked the tripod out of true. Neither camera nor cameraman was injured; yet, with the 200 pounds of bone and frozen flesh throwing up a veritable geyser of pulverized ice and snow at its impact with the glacier, I have always felt that the passing of the missile six inches farther to the right would have torn both to pieces.

The picture, when I saw it on the screen in Montreal, proved most realistic—a highly thrilling and convincing piece of Nature photography!

It was doubtless Harmon's experience on this occasion which was responsible for his decision to confine such animal shots as he made on our present expedition strictly to living specimens, with close-ups of all kind absolutely barred.

Chapter IV

BOW LAKE TO THE SASKATCHEWAN

With the morning of August 20 promising at least a few hours of perfect photographic weather, we proceeded to show our thankfulness by contriving one or two little effects calculated to encourage Nature for providing so nearly an ideal picture setting at Bow Lake and Glacier. I have already described how the towering ice wall, with its flanking waterfalls, ran back in successive waves of *serracs* to and over the crest of the continental device. This was just as it should be for a picture; and so was the many-armed lake, extending from the ice-fronts which gave it life along a red-brown mountain wall to the broken water of its draining rapids. All that was lacking was a proper foreground, and this we hastened to add by pitching a camp on the beach opposite the face of the main glacier.

In regions of savage storms like the Canadian Rockies, campsites must be chosen for utility rather than picturesqueness. Shelter from the winds is the first consideration, and protecting cliffs or timber almost invariably close or restrict the long, unbroken sweep of vista so necessary for the ideal camp shot. And so we took down our tepee and set it up for a few hours just where it was needed to complete the composition of the lakescape—a jutting point of shingle thrown out by the clear mountain streamlet winding down through the flowery meadows from Bow Pass.

That the jauntily cocked pyramid of sticks and canvas was open to every wind that blew had nothing to do with the matter. Neither did the fact that a six-inch rise of the mountain torrent which had laid the tepee's precarious foundation of pebbles would have swept it into the lake. Nor yet did it matter that there was not a stick of firewood within 300 yards. It was pictures we were after today, not shelter or comfort; and for pictures the location of the tepee on that wave-washed and windswept strip of pebbles was almost perfection.

With the camera set up in the middle of the streamlet and turned southwest, the ideally balanced composition included our improvised camp in the foreground, the lake in the middle distance, with Bow Glacier—showing up much as I described it at our first view from the easterly arm—forming a sun-brilliant background against a shifting wall of light-shot clouds.

Set up on the beach and turned southeast, the finder of the camera showed tepee and lake, the rocky pine-clad islets above "The Narrows," and a mountainside ripped and scarred by the savage claws of the bizarre Crowfoot Glacier.[2]

With our "ideal setting" complete and our full battery of cameras in action, we shot the lake and its ice-crowned sentinels, mood by mood, from sunrise to sunset.

There was an hour of repose in the early morning, when every glacier and rock and tree-covered point was reflected, to the last, least detail, in the glassy, unrippling surface. Then there was a spell of smiling under the golden glow of the light-suffused eastern clouds, followed by a madcap dance of mirth and laughter, with the direct sunlight turning the breeze-stirred waves to shoals of diamonds and emeralds.

Laughter gave way to frowning, when shoal on shoal of murky thunderclouds poured over the Continental Divide and quenched the golden flame of the streaming sunshine under a swiftly flung pall of sudden night. And when another wild blade of a freelancing squall came bounding up the valley of the Bow and attacked the first in the middle of the lake, bluster and frown gave way to real tantrums.

No star of the movies ever registered more moods and expressions between daylight and dark than did this lovely Lady of Bow Lake. By dint of much cranking and focussing, we transferred them, one after the other, to imperishable celluloid.

A flare-back of the tantrum mood caught Harmon and me and

the whole flock of cameras in mid-lake in a diminutive homemade boat while I was trying to pull him across to get a close-up of the Crowfoot Glacier. There were a few hectic moments when the conflict of warring airs suggested the classic description of a kindred storm in Drummond's poem:

> De win' she blow from nor'-eas'-wes',—
> De sout' win' she blow too—

With the lake trying to stand on end for a mad minute or two, and a flimsy craft that changed shape every time I laid hard upon the oars, there wasn't much to do but try to keep the so-called bow toward the highest wave of the moment. That, and Harmon's lively baling, sufficed by a margin not quite comfortable.

Nor were our apprehensions entirely on the score of the negative buoyancy of the movie camera. It takes a very warm blooded man not to chill in swimming over a hundred yards in glacier water, and we were a good half mile off shore when the little disturbance kicked up. We had no trouble reaching the beach once the centre of the squall went on about its business.

The lesson in the ways of a wind with a mountain lake came in good time. It prevented us from trying the same kind of argosy on other waters that were broader and deeper and just as cold as that temperamental patch of drippings from Bow Glacier.

As the sky cleared toward the end of the afternoon we prepared to release our first pigeons. The messages, giving the names of the several radio stations already picked up and additions to lists of supplies that were to be sent to us some weeks later at Jasper, were typed compactly on the oiled paper provided for that purpose. After pulling loose several tail feathers in an endeavour to attach the tightly folded slips as directed by experts, we finally abandoned that plan and slipped them inside of the leg-rings worn by each bird to carry its registered number. Tied with silk threads, the tiny rolls appeared in a way to ride quite securely.

Two birds—a pair—were released simultaneously. Teaming up at once, they rose in widening circles for perhaps a thousand feet, and then made off, apparently with great confidence, on a line a bit to the east of the general direction of the valley of the Bow. As this was almost the exact compass bearing of Banff, we felt certain they would be pecking at the door of their home cote inside of a couple of hours.

Just why they failed to fulfil our hopes we never learned, but the chances are that, failing to find—or possibly failing to effect entrance after finding—their former home, they went in search of another. The messages were mailed to Banff a few days later, but whoever found the pigeons evidently thought them too attractive to part with.

The astonishing adaptability of the carrier pigeons to the roughest of travel conditions was a source of never-ending wonder to us throughout the trip. Apparently not the least troubled by the swaying of their box on the top of a pack, nor even by the crushing in or knocking off of their flimsy home by an overhanging limb, they were always ready for their sparsely doled ration of cracked grains and never failed to meet with a caressing peck the friendly finger poked in by way of greeting through a tiny window. We heard them crooning their contentment on nights when the camp was blanketed in snow and on days of grief and drizzle in the sodden flooded flats. They would even chirrup reassuringly to each other when their pack horse was bogged to his eyes, with the next moment threatening their own engulfment in glacial mud.

We found the game little aerial navigators boon companions from first to last, and I never released one of the warm little bodies from my hand to begin its orientating spirals above the icy peaks that separated it from its home cote without a real tug at the heart strings.

Three days' rest with prime grazing had made a great difference in the condition of the horses by the time we broke camp again on the 21st, and, though travel conditions continued no less arduous than before, much steadier progress was made. Bow Pass, a little below timberline, was reached and crossed by an easy grade. The crest of the watershed—in an open meadow thick with lush grass and fragrant with mountain flowers—sloped so gradually in both directions that the divide was barely discernible.

The descent to the valley of the Mistaya—rough, abrupt and slippery though it proved—was effected with the knocking off of only three or four packs. One of these, unfortunately, contained the radio. The pack box of the latter, brass-screwed and padded on the inside, was constructed to withstand rough usage, and that is why it only changed shape instead of flying apart the first time the hulking cayuse carrying it was brought up sharp as his overly broad load jammed between two close-growing pines.

The stout case of tough cedar stood that first collision astonishingly well; also the bufferings of a somersaulting roll down a slide of avalanche-strewn rocks to a temporary resting place in a bower of sylvan beauty where a crystal clear spring bubbled out of the limestone of the opposite wall of the can. It was even recognizable as a box after it had been on the underside of the pack for ten minutes while its bearer sunk to his ears in a bottomless patch of muskeg.

But when a chafed lash-rope parted and let the casket slide down and dangle against "Wolverine's" temperamental heels, the moment had come when it was no longer possible to follow to the letter the admonition of the Banff electrical expert, who had warned us solemnly that the usefulness of the set would depend upon our keeping the batteries hooked up *exactly* as he had assembled them. Even finding the component parts was difficult enough, for "Wolverine" had proved his qualifications as a radio horse by "broadcasting" the battered fragments of the outfit through half a mile of swamp, timber and boulder-covered mountainside.

Abandoning hopes of ever assembling again a wreck which, like Humpty Dumpty, might well have defied the best technical efforts of "all the King's horses and all the King's men," we simply collected such pieces we could find, roped them up in the shattered case and took them along as a potential movie "prop"—something calculated to give a touch of topical interest to the camp shots.

That the battered batch of junk could ever again be coaxed into performing its original function no one but our sanguine French-Canadian cook had the temerity to maintain, and even Ulus' soaring optimism, along with the rest of the outfit, underwent a serious dampening when that containing the remnants of the radio came in for the worst soaking of all the packs, while several of the horses went down in the long boggy ford above the Waterfowl Lakes.

A splendid buck, chased by the dogs, gave us a wonderful exhibition of grace and agility at one of the upper rapids of the Mistaya. Submerged to his horn-tips in a roaring foam white chute at the end of his first jump, he rebounded with the airy grace of Venus emerging from the Cyprian sea froth at his second, to land solidly upon the farther bank of the torrent. "Buster" and "Tip," with lolling tongues, were left gaping foolishly and yelping futile protest in the midst of the muddy seep to which they had skidded in bringing up short as the tumbling cascade yawned below them. They thought they had been running an ordinary four-footed animal, and it had turned out to be something scarcely less elusive than the goose whose whirring wings had baffled them in the Bow.

A herd of a dozen deer, offering easy shots as they watched us file down the valley, were passed unmolested. This was for two reasons. We were still too heavily packed to have room to carry meat, and we were also about to enter a region in which there were few points at which an hour's climb with a rifle would not result in goat or sheep. We had neither the time nor the inclination to shoot for trophies, a pastime that is more and more coming to be restricted to the novice and the tenderfoot. The killing of big game with modern high-power rifles has become so ridiculously easy in all parts of the world as no longer to deserve the name of sport.

Park boundaries, which we touched and crossed at several points, interfered somewhat at first with our shooting for meat at the moment a shortage occurred. As soon as pack room was available, however, it was easy to lay in enough of a supply in the unrestricted sections to carry us through those in which shooting was not legal.

Meat keeps a long time at those high cold altitudes, and after the first week we were rarely out of it during the whole trip.

The Mistaya was a roaring boulder-strewn torrent where we first came down to it, but at the end of a mile broadened out into meandering channels emptying into the head of Upper Waterfowl Lake. Several deep fordings and a long, wet wade through boggy marshes took us to a precarious camping ground among the burned timber on the sloping mountainside east of the head of the lake. The latter, beautiful to the eye, especially from a high altitude, was too boggy around the border to permit even the dipping of drinking water without risk of being mired.

We remained over here for a day to climb a couple of thousand feet to the summit of the easterly ridge for a vantage from which to take pictures of the splendid panorama of Howse Peak and Pyramid, with the silvered slivers of the Upper and Lower Waterfowl Lakes, harnessed in tandem by the gleaming ribbons of the Mistaya's channels, prancing in the sunlit foreground.

As it was impossible to ford the Mistaya in the canyons below, the only practicable route on to the Saskatchewan was to retrace our steps around the head of the lakes and go down the western side. This entailed deep but not especially troublesome fording; or, rather, it would not have been troublesome had the horses had the sense to keep to the crossings into which they were headed.

Among the animals which thought they knew a better way was the young mare carrying the salt and sugar. Rolled head under at the little riffle below the ford and carried down a couple of hundred feet before pawing onto solid footing, the venturesome filly trotted jauntily out, smeared herself with muck by a roll in the nearest mudhole, and then galloped on to set the pace for the pack train all the rest of the morning.

Comprehension of the reason for the culprit's blitheness of spirit came when we unpacked at the end of the afternoon. What with leakage of brine from the salt sacks and syrup from the sugar, Nelly had reduced the weight of her pack at the rate of rather better than ten pounds to the mile all the way to the Saskatchewan.

This must have been the sort of thing which "Soapy" had in mind when he spoke of letting Nature take her course in the matter of reducing packs. To do the veteran justice, he seemed as much upset as any of us over the loss, and even promised to punish a possible repetition of the offence by giving Nelly the dried fruit and dehydrated vegetables to carry when we took the trail again. This he forgot to do, however, much to our ultimate sorrow.

We came down to the Saskatchewan a mile above where it receives the Mistaya in the shadow of pinnacled Mount Murchison, and just below the junction of the North Fork and Howse River.

Although not over from 15 to 25 miles from its principal glacial sources, it is already a mighty river, varying in width from 500 yards to half a mile according to the stage.

Nothing allows a river to accumulate volume so rapidly as a series of great glacial reservoirs tapped by its head waters, and in few if any of the great mountain system of the world are these ice feeders located so favourably for the rapid augmentation of flow as in the Columbia Icefield region.

The Amazon, which probably discharges 50 times the amount of water into the Atlantic as does the Saskatchewan into Hudson's Bay, almost certainly has no tributary with a flow a hundred miles from the summits of the Cordillera of the Andes equal to that of the Canadian river 25 miles from the Columbia Icefield. I am also inclined to think that the same would hold true in a general way of the Yangtse, flowing east from the Tibetan plateau, and the Indus and Brahmaputra, flowing south.

The Athabasca, flowing north from the Columbia Icefield, probably has an even greater volume in its upper reaches than has the Saskatchewan. On the other hand, the drainage to the Columbia, by the Wood and Bush rivers, is much smaller.

Working cautiously from bar to bar and covering perhaps 700 yards of quick-flowing, but not dangerous water, we reached the north bank without swimming the horses or seriously wetting a pack.

Camp was pitched in the timber at a magnificent point of

vantage just below the mouth of the North Fork. That the place was an old Indian rendezvous was indicated by rotting tepee poles, the bent willow frames of steam bathhouses, and deep layers of musty hair scraped by the tanners from hides of deer, elk and moose.

The Indians of the plains never established permanent habitations far inside the Rockies, but that they hunted the vast system to the very rim of the ice caps of the Continental Divide was indicated by their ancient and long disused trails, which we found on every pass and often far down into the timber.

Never breaking down the earth save by the soft pat of moccasins or by the pressure of unshod hoofs, and rarely if ever cutting out timber, fallen or standing, these old hunting parties still left behind them fragments of trails that could be improved only by the liberal use of axe and pick. Their grades are invariably as favourable as the topography permits.

Not satisfied either with the lighting or the backgrounds for his original motion picture of the pack train crossing the Saskatchewan, Harmon decided to wait over and make another under better conditions. Two days of warm, cloudy weather were unfavourable for pictures but did have the effect of starting one of the heaviest late summer rises that the Rockies have known in many years.

The river, fed from its many melting glaciers, had risen two feet and probably more than doubled in volume when the third day brought bright skies and a flood of hot sunshine. With the crossing picked for the pictorial possibilities rather than for its facilities as a ford, it was evident at the outset that we were in for a lively time in taking the pack train over the swollen river. For this reason dummy packs of canvas-wrapped fir boughs were substituted for boxes, bedrolls and the regular run of camp stuff. This plan had the double advantage of giving the horses lighter loads and of effectually precluding the possibility of a further soaking of outfit and provisions. There had been so great an increase in the volume of the river that Harmon and La Casse succeeded only on the second attempt in crossing with the movie outfit at the broad, shallow ford which we had waded with so little trouble three days previously. Setting up the camera on a high bench on the south bank, just above a point where almost the whole flow of the Saskatchewan was concentrated into a single 200-yard-wide channel, they signalled for us to come on with the pack train.

"Soapy" led the way, with Baptie and me urging on the huddled, reluctant horses from the rear. "Buster," our collie-husky crossbreed, rode one of the packs. "Tip," our camp-following Indian mongrel, who had funked a similar seat, I carried under my free arm.

"Soapy's" mare was swimming the moment she stepped off the gravel bar along the shore; likewise the rest of the animals as they followed suit. Several of them were ducked head-under as they lost their footings, to come up sputtering and wild-eyed, with each eager to climb to safety on the back of its neighbour.

Packs were badly bumped and jostled, with two or three bundles of young Christmas trees breaking loose and floating downstream in mute vindication of the wisdom of substituting them for the sugar and the radio. "Buster" was knocked from his seat at the first souse; but "Tip," clinging with an almost catlike grip, managed to keep hooked on to the crook of my arm.

Orientating themselves as the shock of the first plunge passed, the horses spread out and began lunging along in the wake of the leader, which "Soapy" was trying hard to keep swimming at an angle that would carry her out at the only practicable landing on the cut and broken opposite bank.

Seeing the swift current was setting him down much faster than had been calculated, with the alternative of making the landing or going onto, and possibly under, a pile of drift logs blocking the entrance of a side channel opening just below, he slid off backwards and towed by the tail to give his horse a better chance. Still failing to close with the landing fast enough, "Soapy" wisely headed his horse back to midstream so as to give the drift a wide berth and make a bar where the river broadened and shallowed below.

I would have done better to follow "Soapy's" lead while there was time and room to avoid the treacherous bank. With my powerful mount swimming strongly and with plenty of reserve strength, however, I was reluctant to forego the chance to make the landing as prearranged and so keep all the pack animals that would follow as near the camera as possible.

We made the bank, but too far down by a scant yard or two. The caving earth broke back under pawing hoofs, and both "Jerry" and I were doused well below the foam-flecked surface when we tumbled back into the icy current.

The shifty "Tip," sensing his chance, jumped at exactly the right moment, landing on dry ground with an almost dry hide.

I slipped clear of the saddle to give my mount a better chance to recover his balance, but climbed back with alacrity on discovering that all of the immediately adjacent river was filled with floundering and more or less upended pack horses, all in difficulties similar to those of Jerry and myself, and for the same reason.

The next moment the whole mob of us was slapped by the ten-mile current against the jutting jam of logs.

With each of us trying to kick and climb over his wallowing, snorting neighbour, it was exceptionally good luck that none of the various participants in the melee were much banged up. By really good fortune there was not enough water drawing beneath the logs to create a dangerous under-suck, nor yet enough flowing over them to tempt the horses to risk broken legs by climbing the obstruction. After we had pawed and jostled each other for a lively minute or so, the current got sufficiently behind the milling mob to roll it around the end of the barrier, from where it was straight swimming to the bar below.

Seeing the high-class action that was going to waste under the high bank, Harmon brought his long-distance shot to a sudden end and rushed down for a close-up. Setting up with feverish haste, he was just in time for the finale at the logjam.

There was a twinkle in his eye when, a few minutes later, he sauntered up to where we were drying out by tripping an Indian dance round a log fire and casually asked if we'd mind doing the first part of that close-up over again! "Soapy" replied with a guffaw behalf of the horses, while I made similar response on my own account.

Harmon admitted that the request was not made seriously, and that he fully understood without trying it personally that a drift pile throwing off a ten-mile-an-hour current was no place to take a pack train, even one carrying nothing more valuable than Christmas trees. We did make the return by swimming the same channel, however, but starting high enough up to work well clear of the cut bank and the jutting log pile.

While filming the rough, steep gorge of the North Fork the following morning we were presented with the opportunity to make what turned out to be one of our best wild animal shots. A black bear playing with her cubs on a patch of sunlit rocks was the subject. After a surreptitious shooting of the amusing antics of the trio with his long-focus lens from a comfortable distance, Harmon suggested that it might bring about an interesting variation of action if I would close in and make our presence known by some such friendly action as throwing a stone or giving a lusty yell. I chose the latter as the less belligerent means of creating the desired diversion.

The effect of my Apache war whoop, delivered from cupped hands at not over 50 feet from the three pairs of sharp, backlaid ears, was all that could have been wished for. Without even looking to see where the blast came from, that capable old mother swatted swiftly at a cub with either paw, and then herded them pell-mell up a nearby Jack pine.

The instant the hindmost of the precious pair was clear of the ground, round she wheeled and came charging back to settle with the enemy. Full tilt she came, right to the brink of the gorge, and there, perforce, she was brought to an abrupt and sliding stop.

That 50-foot-wide canyon of the North Fork of the Saskatchewan was the keystone of our strategy, for we and the camera were on one side of it and the temperamental Mother Bruin on the other. That is quite the best way to arrange bear pictures in the open, especially when guns have been left in camp. The film turned out quite perfect to the last detail, even showing the cubs peeking round the sheltering tree to watch their mother shake an admonitory paw at us.

Chapter v

BUCKING MUD AND FLOOD

We had our first serious trouble with far-straying horses during the three days spent in camp at the forks of the Saskatchewan. This was probably due to the fact that at least three or four of the animals knew exactly where they were, and that the rich grass of the fertile Kootenay Plains—a broadening of the lower river valley—was not many hours away.

Rob, the wrangler, who knew no peace of mind when a single unit of his bunch was unaccounted for, spent many hours each day turning back and rounding up the runaways. Using hobbles in a land of burned and fallen timber entailed too much risk of "hanging up" a horse over a log; indeed, my own powerful and spirited mare, La Belle, which "Soapy" had conditioned especially to carry the 240 pounds which I weighed at the beginning of the trip, snagged herself seriously in the belly with all four legs free.

A Chinook wind from the Pacific, warm and soft as new milk, had been playing on the icefields to the north and west during all of the 60 hours we remained in camp on the bench above the junction of the two main branches of the Saskatchewan. When we took the trail again on the morning of August 26 the humid spell was over and a brilliant flood of yellow sunshine was pouring from a deep turquoise sky unflecked by cloud or undimmed by haze to the farthest segment of peak-notched skyline.

There was a diamond glitter on the lofty pinnacle of Forbes and the serrated line of its sister peaks along the Continental Divide, while near and far every mountainside was streaked with the "downward smoke" of waterfalls, shimmering brocades of jewelled silk in the sunshine, flutters of wind-blown carded wool in the shadows.

The Saskatchewan was out of its banks at the forks, spreading over the flats and encroaching upon the high mark of the spring floods. In the narrow gorge of the North Fork, where the water was up from 20 to 25 feet in three days, what had been cascades and abrupt falls of six feet and more were completely wiped out by a solid white stretch of tumbling torrent.

As long as the horses could find footings on the rocky lower slopes of the long massif of Mount Wilson which bulwarks the North Fork to the east, we were able to avoid the spreading floods on the flats. When talus and snowslide débris finally presented unsurmountable barriers, there was no alternative but to push on up the valley along the inundated bottoms.

Then it was we learned the meaning of real water trouble. Although successful in avoiding a complete ford of the main river, there were the endless networks of back channels to be passed, many of them deep and steep banked. With the surging water practically opaque, there was no telling whether the next step was going to land a horse to his fetlocks or up to his mane.

It was here that we paid the penalty for the false courage bred in the horses as a consequence of our little exhibition for the movie with the dummy packs. Up to that time nearly all of them had been extremely nervous about tackling a deep ford; afterwards they had altogether too much confidence. Like the advocates of inland canal extension, theirs might have been the slogan, "The water way is the cheapest and best." Let the road we had blazed for them be rocky, boggy, or blocked with deadfalls, and forthwith the three or four leading spirits among them would bolt toward supposed easier going in the nearest water, be it the boulder bed of a cascade or the cutbank of a slough.

After a few days we came to understand and to anticipate these outbreaks toward the new freedom; but that first morning on the North Fork of the Saskatchewan we were caught quite unprepared.

Without the blink of an eye, Nelly, the sugar and salt horse, walked off a bank of glacial silt that was caving and receding before the attack of the grey-green torrent of the main river. Rolled over and whisked away in an instant, she came up with a blithe snort and started swimming straight for Hudson's Bay.

What with the 12-mile current and her purposeful pawing, she travelled at five times the speed at which we could force our horses through the brush and mud of the flooded flats, and was out of sight around the next bend before we were well started.

Of course she had to strike bottom in time, but it was only by the rarest of luck that the intercepting gravel bar, a quarter of a mile below, chanced to run out from our side of the river. Even here the perverse filly was in two minds about rejoining us, stoutly declining to move a step shoreward until La Casse waded out to bring her back.

All the sugar and salt salvaged from this later and longer baptism was in the shape of the dirty brown chunks which resulted from the reduction by axe of the indurated slabs left after the containing sacks had been dried all night by the fire.

Several other packs were badly soaked as a consequence of plunges into deep water, nearly all of which appeared to be due to the newborn mania of the horses to cure their trail troubles by hydropathic treatment.

The radio was the worst casualty, undergoing a complete submergence when the horse carrying it stepped off into the river and was carried down under the horizontal trunks of several undermined pine trees before he found a place to climb out.

As the radio, due to previous disintegrative bangings and bumpings, was already rated a total loss, we were less concerned over its wetting than about that of the sugar and salt. It was now inevitable that even the miserable remnants of the latter must be exhausted weeks before replenishment would be possible.

What could hardly have been other than a disastrous attempt to ford the main river was avoided by a rough and precarious climb over a jutting headland, where two or three men to a horse were necessary at one point to keep the animals from sliding back to the valley. We pitched camp opposite the Alexandra late in the afternoon, having made about eight miles. Several of the horses were so played out as to be unable to stand until their packs were removed.

The human element of the party was holding up fairly well physically, but the mental atmosphere of the camp was well reflected in a note which "Soapy" wrote, to be left on a forked stick for his partner who was expected to follow later with a hunting party. Slightly expurgated, this missive concluded as follows:

> "If you need any sugar or salt, dip it out of the Saskatch. Nine-tenths of ours is already in the drink and the rest most likely goes tomorrow. If you see any horseshoes floating down stream, look under them for my cayuses. Deep water navigation by pack train ain't what it's cracked up to be."

The main North Fork and the Alexandra must be about equal in average volume. The former continues north for some miles before bending to the west to its source under the Saskatchewan Glacier, itself a tentacle of the Columbia Icefield. The Alexandra flows directly from the west, deriving its waters about equally from the Mount Lyell Icefield and that of the Columbia. Alexandra Glacier, leading up to the Continental Divide under the peak of the same name, was partly visible from our camp on the North Fork, but although little over eight miles distant, we were to be two days in reaching it.

The camp at the junction of the Alexandra and North Fork bears the gruesome but not unfitting name of "The Graveyard." It is located at the focus of a four-way convergence of important mountain routes. That to the south—the one by which we had come—leads by the North Fork, Mistaya and Bow to Lake Louise. That up the North Fork leads finally to Jasper by alternative trails over Nigel or Wilcox Passes. What is for considerable distances hardly more than a blazed trail today follows the route of the projected Banff to Jasper highway, which, when constructed, will be one of the finest scenic roads that can ever be built anywhere in the world.

The westward route—it could hardly fairly be called a trail—from "The Graveyard" runs up the valley of the Alexandra to the Castleguard branch of the river, to come to a blind end against the Saskatchewan Glacier tentacles of the Columbia Icefield. This was the way we planned to follow as far as it went, and then endeavour to go directly across the icefield on a shortcut to the head of the Sunwapta and the Arctic slope of the divide.

The easterly trail from "The Graveyard" winds over the mountain ridge to Pinto Lake and a junction with the trail leading down the Cline to the lower Saskatchewan and on to Banff. This latter trail, winter snows permitting, was to be a part of our return route to the south.

Due to its strategic location, the camp at the mouth of the Alexandra had been a hunting rendezvous for many years, first for the Indians and later for the farthest faring of the parties from Banff or Jasper. Skinning game and discarding unsatisfactory specimens resulted in the accumulation of many bones and heads, and these grisly heaps of unwanted trophies of the chase give point to the name of "Graveyard."

Unsuccessful hunters are occasionally charged with having salvaged from "The Graveyard's" varied stock of discards the trophies denied to their erring rifles. I have seen worse heads proudly displayed in trophy rooms than some of the specimens thrown away to bleach on the flood-scarred flats of the North Fork of the Saskatchewan.

Another humid night, followed by a clear, hot morning, brought still higher water. La Casse, indeed, was inclined to believe that both North Fork and Alexandra were at higher stages than when he had gone over the same route the previous year when the spring thaw was culminating and the rivers were near the crest of the June rise. Further damage to provisions was inevitable in the deep, swift fords ahead, but this prospective loss was far less serious than the threat to the considerable part of Harmon's photographic supplies, for which no watertight containers had been provided. In all of the veteran's 20 years of pack-train travel in the Rockies he had never found it necessary to take special precautions to protect his photographic supplies from water. Now the lack of such protection threatened seriously to jeopardize the success of the expedition. With the movie films and my own roll films in cans, the principal concern was over the hundred or more packets of Harmon's special film-pack and the large motion picture machine. The compact little "Sept" movie camera, with its 16 feet of film which ran through at the release of a spring, rode with Harmon on his horse to be ready for emergency shots. Most of our "still" cameras were also carried where they could be given a certain amount of personal attention in case of trouble.

The best we could do for the threatened camera and supplies was to give their respective pack boxes extra wrappings of canvas before pushing on to breast the flood that was sweeping down the valley of the Alexandra. Provisions and bedding would have to take their chances.

We crossed the score of scattering channels of the North Fork without trouble. The declivity of the valley had increased rapidly within the last two miles, so that the swift flowing water carried all of the light glacial silt down to deposit it in the opener reaches above the box canyon where we had photographed the bears. This made for many broad, shallow channels with hard, rocky bottoms, quite simple and safe to ford.

At the mouth of the Alexandra conditions quite the reverse were encountered. Here the floods had risen to cover the low flats in one unbroken lake, with little to indicate the course of the deep, perpendicular-banked, meandering channels by which it was traversed. After bogging the horses repeatedly in attempting to work along the base of the slope where the waters of the lake lapped the mountainside, "Soapy" decided to try to avoid the flooded area entirely by taking the pack train up through the timber. This led us into quite the roughest and most punishing going we had yet faced.

With a rocky, broken slope of all of 40 degrees to traverse, progress would have been quite difficult enough without any additional natural obstacles. It was the timber which finally took the heart out of us, man and horse. This had been burned in patches. The trees of the unburned stretches stood thickly enough to hang up a pack now and then but still permitted a slow but fairly steady advance. The passage of the tongues of charred slope where the fires had swept presented a far more serious problem.

With the bristling young growth standing thick as the spines on the back of a ruffled porcupine, there was really only one safe and satisfactory way of taking the pack train through it. That was to cut out a swath with an axe. This was slow—especially where the deadfalls laid their trickiest traps underfoot—but it was also sure.

If the horses had been content to wait and let us turn to and systematically clear a way ahead for them, things would not have been so bad. It might have taken the whole day to do it, but we would ultimately have brought them through to the solider flats above the overflow lake without further boggings. Unfortunately, however, action had already become too hectic to make a cool, deliberate analysis of the situation possible on the part of either horses or packers.

With pack covers and tempers—both human and equine—torn to tatters at the end of the first hundred yards of arboreal tunnelling, men and horses said and did things to each other which effectually precluded further peaceful and dispassionate consideration of the situation. The decision in such arguments as were indulged in went mostly to "Wolverine" and the two or three other Indian cayuses which drove home their points with steel-shod hoofs. Even a packer hasn't much to say to that kind of repartee, especially when the horse makes his point first.

"Soapy" voiced his protests in two languages and twice as many Indian dialects—until Nelly, the salt- and sugar-dissolver, with her forelegs clasped in the embrace of a pair of locked deadfalls, lashed out with her hind hoofs and planted a love tap a few inches below the nerve centres of the old woodsman's diaphragm. "Soapy" was quiet for some minutes after that caress, doubtless musing on the ingratitude of a colt which he had raised on a bottle after its mother had been lost in an ice-choked ford of the lower Saskatchewan four years previously.

None of the rest of us had to endure the mental anguish of being kicked by his own bottle-baby, but that didn't reduce by a whit the physical discomfort of having a horseshoe clapped over one's ear just as he was bending to disentangle the lash-rope of a scattered pack. And we used to think it was lucky to find a horseshoe! Doubtless it all depends upon how and where you find it. I can only testify that as an earmuff the luck attaching to a horseshoe is of a distinctly negative character.

It was a number of little incidents of the kind indicated which made difficult, and finally impossible, the co-operation between horse and man imperative for the overcoming of the almost prohibitive obstacles incident to taking the pack train any great distance along that steeply sloping, heavily timbered mountainside. Sixteen horses starring in as many different directions, with only five men to follow them, did not make for linear progress, and the difficulty of rounding them up was considerably hampered by the fact almost every wake was strewn with fragments of broken packs. Things like the contents of smashed grub boxes and a snag-ripped sack of rolled oats take a lot of salvaging in soggy foot-deep moss and fallen timber.

Constantly reslinging packs and dragging fugitive horses back into line, we had made scarcely more than a mile by early afternoon. To reach the nearest practicable camping ground before dark, therefore, left no other alternative than to drop back to the flooded flats and see what could be done in floundering across them.

There was still an unbroken sheet of water stretching from wall to wall of the valley where we came out of the timber into the open again, but "Soapy," shouting optimistically that the lake was much shallower than below, plunged boldly in to show the pack train how the thing should be done. Possibly he was correct as to the soundings of an average cross-section right across the valley, but I have serious doubts if there was a profounder pool in all of the Alexandra than that into which our doughty old leader pushed his lithe-limbed thoroughbred at the initial plunge.

Horse and man disappeared completely from sight and it was all of two or three seconds before anything but bubbles and swirls marked the point of engulfment. From the fact that mount preceded rider back into the sunlight, I assumed that the hole had been deep enough to allow both of them to roll over at least once without flicking spur or horseshoe above the surface. The main surge of the river appeared to have undermined the root-bound bank to a depth of ten or 15 feet, and "Soapy" and his horse, under the impulse of their rush from above, had dived most of the way to the bottom. A horse is a very buoyant animal. On very few occasions have I seen one completely submerged for so long. "Soapy's" Rooseveltian touch had gone with his glasses, but, with two inches additional droop to the sheriffian moustachios, he was more than ever the tusked bull walrus. There was little of the basso profundo roar of the walrus, however, in the throaty croak with which he accompanied the pantomime intended to convey to us that, along with a lot of glacial water and mud, he had swallowed his "dining room set."

When Rob explained that "dining room set" was "Soapy's" facetious euphemism for his false teeth, Harmon and I choked back our ill-timed mirth and began forthwith applying vigorous first aid in the form of lusty slaps on the sufferer's buckskin-shirted back. Fortunately, only the mud and water had taken a through passage. The teeth, hung up somewhere in the upper reaches of "Soapy's" œsophagus, finally were dislodged by the coughing induced by the slaps and gradually jiggered along back to where they belonged.

Two horses which had plunged in after the leader swam on to a firm footing on the flats, but only after both of their packs had been doused completely under. As continuing on this course this would have meant more damaged provisions, to say nothing of the risk to cameras and photographic supplies, we decided to try to skirt the margin of the overflow for a few hundred yards farther. This was accompanied by several more deep and troublesome boggings, but we continued to wallow along toward a point at which it appeared practicable finally to begin wading the flats.

The last 50 yards was over a sharply sloping shelf of broken limestone, which was slippery with moss and streaked with seepage from the mountainside. Horse after horse lost its footing in the helter-skelter scramble across, but none of the first dozen went sufficiently out of control to do a tailspin or nosedive.

When my own turn came the exercise of ordinary common sense would have warned me that the proper thing to do was to dismount and give my horse a fairer chance by leading him over the most treacherous part of the slippery ledge. The animal which I was riding temporarily in place of the surefooted "La Belle," who was still too sore from her snagging to carry a saddle, was one of those secured at Lake Louise to replace the two strays of the original pack train. He was a powerful and willing brute but handicapped by a splay hoof and a terpsichorean temperament that inclined him to spells of toe-dancing at highly inopportune places and occasions. It was the bad hoof which started the trouble at the sloping shelf. When "Le Diable's" forefeet commenced slipping he tried to remedy the difficulty by rearing them high in the air and walking on his hind ones. As one of the latter carried the deformed hoof, stability, far from being improved, became a deal more precarious. In fact, "Le Diable," after a clattering spell of buck-and-winging with his steel-shod hoofs striking sparks from the rock, started to topple over backward.

I kicked free of the stirrups as I felt him going and, more by luck than calculated intent, struck on my back on the upper side of the shelf. The impact jolted me all the way up the spine, but this was of small moment in view of my good fortune in landing in a niche which prevented my jarred anatomy from sliding down to interfere with the highly intricate *pas seul* by which "Le Diable" was expressing his elation over the fact that he had rid himself of a rider without losing his own balance.

His triumph was short lived. When the splay hoof skidded on

a patch of dewy moss "Le Diable," like Lucifer, came out of the skies with a bang. Caroming from the ledge against a Jack pine, he was flung back to the rock again, this time with all four hoofs pawing the air. And it was in this ignominious posture that the prideful devil-dancer of a moment before slid down to and stuck fast in the morass of mud and burned timber at the edge of the overflow.

A bit dazed from the jolt, I was still able to scramble to my feet and lend a hand to the packers in extricating my mount from his difficulties. A sore spot between the shoulder blades, where the box of my camera had interposed to break an otherwise even contact with the face of the ledge, was my only souvenir of the occasion. "Le Diable" had gone farther and fared considerably worse. Sizable patches of hair and hide were replaced by raw abrasion at several exposed points, while the slide down the rock with the saddle beneath him appeared to have strained his powerful back. By good fortune opportunity offered to replace him with a more dependable animal before the really serious work of the trip began at the Columbia Icefield.

At a point near the head of the overflow lake we were finally able to venture out onto the flats without great risk of wetting the packs except as a consequence of untoward accident. For a mile we splashed along through gradually shallowing water, finally to come out upon a stretch of muddy but submerged meadows cut with many meandering channels of the river. Depth of water, rather than swiftness, was the menace here, but by exploring carefully in advance with the saddle animals, serious trouble with the packs was avoided.

Although we were little troubled with deep bogging of the horses after the increasing slope of the valley floor brought better drainage of the stretches subject to inundation, progress was still terribly tiring upon the heavily overloaded animals. Even the solidest of the glacial silt tended to form a vacuum cup under each hoof, the breaking of the suction of which demanded a physical effort probably greater than that of climbing a steep trail. For the first time since our departure the horses became so exhausted as to try lying down with their loads. As this always slacked the lash-ropes, it usually proved easier to throw off the packs entirely before dragging the wearied animals again to their feet. And, that, of course, meant five or ten minutes delay for the whole train for every reslinging of a hitch.

With no possible camping ground offered either by the steep mountainsides or the muddy flats, there was nothing to do but keep pushing on to where the rough, sharply sloping boulder-fan from a torrent draining a glacier to the south poured down to the Alexandra. It was a sodden, ding-dong struggle to the last foot, and the mauve shadows of coming night were banking thickly below the wall of the imminent Continental Divide before the last of the straggling horses scrambled up out of the mud to a solid footing of shale and gravel.

There was no grass and little protection from the wind on the forbidding triangle of torrent-strewn rocks eroded from the southern valley wall, but with no other possible halting place available it was up to us to make the best of what we had. Fairly good forage was provided by swimming the unpacked horses across a back-channel of the river and turning them loose on the half-submerged flats beyond. Tepee- and tent-poles were found after extended and exhausting cruising of the scantily timbered mountainside.

With no evidences of anyone ever having camped at this point before, "Soapy," his interest in the lighter and finer things of life reviving as La Casse began to spread the supper dishes, announced that it was in order for us to give the site a name. Even the fumes of steaming coffee and frying ham could not quite obliterate the memories of that accursed day of bangings and wallowings, and that, of course, made it impossible for us to take up the matter impartially and without passion. Of the several names proposed, only the one advanced by the gentle Harmon was entirely fit to print, and even that was not quite polite.

It is wisely provided by the American and Canadian Governments that official geographical nomenclature shall be vested in boards sitting far away and long after the event of discovery. If the names applied to muddy rivers by the pioneers were perpetuated, the impression might well be fostered that every stream in question was a tributary of the Styx.

Fording a side channel and crossing half a mile of overflowed flats after breaking camp the next morning, we skirted another delta of gravel to come out upon a stretch of valley of greatly increased declivity. The river was swifter and narrower here, with broad, hard gravel bars between the winding channels. It was really much less exhausting going than that of the previous day, but the horses, weak from overwork and underfeeding, had much trouble at the ever recurring fords. At the end of two miles all of them were straggling badly, with several persisting in lying down with their packs.

It was at this untimely junction that there opened up a magnificent vista of valley, closed at the far end by the sun-sparkling *serracs* of Alexandria Glacier running back to the snowy summit of the peak of the same name at the crest of the Continental Divide. The setting was an incomparable one for the series of fording shots which Harmon had been postponing until he found just the place he wanted for them.

It would not have been so bad had the layout of the scenery been such as to permit of the crossings being made at the most favourable points in the regular way. These were hard enough for the jaded horses at the best; when exigencies of lighting background made it necessary for them to be put in and driven through at bends where deep, swift channels running under steep, abrupt cutbanks rendered it difficult to scramble out, the temptation to challenge the right of the movie to interfere with the regular routine of trail work must have been a serious one for "Soapy" to withstand.

Not a little worried for the last two days over the way in which the abnormally severe work was wearing down his horses before the trip was well started, it must have struck the old woodsman as a bit hard to have additional effort of the most arduous kind demanded of animals already near exhaustion. Harmon had faced the same sort of problem many times in previous seasons, and, therefore, had had the foresight to explain it to the packers in advance. Thus forewarned, old "Soapy" gamely came through with the active and hearty cooperation without which the desired shots could not have been made. This made it not any the less trying to have to see all-but-collapsed horses rolled against a logjam and forced to do an extra hundred yards of swimming and floundering just below a broad, open, hard-bottomed ford.

The horses carrying the cameras and photographic supplies were headed over at the safest crossings. The packs of most of the other animals came in for thorough soakings. Two or three bedrolls took in water, as did also the dried fruit and dehydrated vegetables. More brine and syrup streamed down the river from the salt and sugar sacks. All of which was of very little moment, however, when weighed against Harmon's triumphant announcement that "the backlighting across Alexandra shot like a million dollars."

One of the longest and hardest fords of the day furnished an interesting and not altogether explicable instance of canine psychology. The pack train made the crossing without getting into serious difficulty but the hard, partly upstream swim in swift, broken water was almost too much for the dogs. The powerfully muscled "Buster" made it at his first attempt, but only by a hair. "Tip" failed by a good five feet.

Carried down a hundred yards and back to the side from which he had started, the little Indian mongrel promptly galloped up the bank and put in again at the exact point at which he had made his first start. Failure this time was by a wider margin. Twice more he repeated the attempt, each time with a wider chute of swirling water separating the point at which he began to lose ground from the striven-for bank.

With tottering legs and lolling tongue, the game little beast dragged his tired body back out of the icy waters after his fourth failure. His strength was plainly nearly gone, but something inside of the funny flat-topped head was only beginning to come into action. It could not have been reason, for "Tip" had been too busy keeping his head above water to have time for any real thinking even had he been capable of it. Instinct, perhaps, is the more convenient word; but even that somewhat inclusive term does not quite satisfy.

At any rate, without an instant's pause—and just as I had resigned myself to recrossing and bringing him back under my arm—"Tip" began loping up the bank again. Passing without pause or side-glance the point at which he had launched his four previous swims, he ran on a hundred yards to disappear in the timber of the mountainside beyond a bend. Recurrent flashes of brown fur between the trunks of the trees revealed him running on to where a gravelly beach marked the beginning of a ford so broad and shallow that half of it was passed by long bounds.

Attribute it to "the unerring needle of animal instinct" if you will; but what, then, was that needle doing at the first four abortive swims?

"Buster," everything considered, impressed me as being quite the most intelligent animal of any kind I had ever had opportunity to study at close range. And "Tip" I have often thought of as one of the worst fools of a dog that ever came to my notice. Yet that confounded camp-robbing Indian mongrel revealed occasional flashes of intelligence, instinct, or what you will, which not only quite out-Bustered "Buster," but even left the several human units of the party blinking blankly at each other with wondering, uncomprehending eyes. One such instance occurred in connection with our discovery of the great spring feeding the cataract on Castleguard River; another on the Athabasca Glacier. I will tell of both in their proper sequence.

After making a total of not over four miles for the day, horses were unpacked and camp pitched at a beautiful and well-sheltered site a mile below the foot of the Alexandra Glacier. Several peaks of the lofty massif of Mount Lyell were visible to the southward, rising above a broken but very extensive icefield. This great *mer de glace*, discovered by Dr. Hector of the Palisser Expedition in 1858, has a total area of nearly 40 square miles. Until the discovery of the Columbia Icefield, 40 years later, it was believed to be the largest in the Canadian Rockies. All five of the Lyell peaks are over 11,000 feet in height, while the summit of Forbes, a short distance to the south, attains an altitude of 11,902 feet.

Clambering over a low ridge with the movie outfit in search of a vantage from which to make a picture of the Alexandra Glacier, we stumbled upon a set of tepee poles marking the site of a comparatively recent Indian camp. "Soapy" promptly announced that we were viewing the remains of an Indian's "Honeymoon lodge," going on to explain that it was purposely pitched at a distance from water in order that the bride should have opportunity to show her quality as a worker at the outset.

"Soapy," opining that there was much to be said in favour of the Indian custom of not spoiling a bride by soft pampering, demanded point-blank my own views on the subject.

I replied "Ugh!" this being as near as I could come in Cree to, "That all depends upon the bride." I can conceive of nothing less wise than committing oneself any further than that upon so delicate a subject, especially in print.

Returning to camp just before dark, we found another outfit had arrived. It was that of Dr. Fowler, of New York, accompanied by his son and Dr. Atkins, of Banff. Bill Potts, "Soapy's" partner, was head packer. With him as wrangler was an extraordinarily tall young American, aptly described by "Soapy" as being "as long as a lash-rope."

Dr. Fowler was a Canadian Rockies "pilgrim" of many years standing. He did no hunting or climbing but was very keen on colour photography. He planned to follow our route to Castleguard Valley and the Columbia Icefield, backtracking later to the North Fork of the Saskatchewan and going on to Jasper by the regular route over Nigel Pass.

Having long heard of Bill Potts as one of the doughtiest of the big game hunters of the Rockies, the present opportunity to get him and old "Soapy" together, matching yarn for yarn, was too good to be missed. Seeing the shadows of the two famous woodsmen bobbing together against the roof of the Potts cook tent, I hied over hotfoot, hopeful of hearing at first-hand Bill's story of the time he had swum the swollen Brazeau with a grizzly cub in his mouth after braining its mother with an axe.

A low mutter of conversation became audible as I approached, with a steady clickety-clicking obligato running through it. Lifting the flap of the tent, I was just in time to hear "Soapy" assure Bill that "Frog" La Casse had "the darndest slickest receep" for a frosting for writing letters on "choklit cake" that anyone had ever heard tell of. And the clickety-clicking was from two hard-plied pairs of knitting-needles!

When "Soapy" resumed the rhapsody interrupted by my entrance, it was to offer to trade Potts "The Frog's" frosting "receep" in return for instruction in the esoteric intricacies of the new "hook-stitch." Those two hard-boiled, hard-bitten old bear-killers had actually settled down to spend the evening knitting bedroom slippers and swapping cooking recipes!

Potts explained to me about the knitting presently. He had taken to it to kill time during a dreary two years in a German prison camp, subsequently passing on the accomplishment to "Soapy" while they hibernated through the long winters at their Morley ranch. But I never did get the proper version of that cub-in-the-mouth swimming episode.

Chapter VI

IN CAMP AT CASTLEGUARD

The night of August 28, spent in the camp below Alexandra Glacier, brought a distinct change of weather. There had been no rain or snow for a week, the first part of that period having been warm and humid, the rest hot and clear. Now it had turned sharp and cold, with a lowering pall of grey clouds threatening a heavy storm.

The promised diversion was a welcome one, provided only that it did not last too long. Forest fires are smouldering all through the summer in the Canadian Rockies, never being more than partially quenched by the rains. This means that a few days of weather without moisture in one form or another inevitably results in that bane of the scenic photographer—smoke.

Only the smoke clouds from a nearby fire, or the dense blanket of a general conflagration, seriously interfere with close-range work, like trail and camp shots. But for strictly scenic work, where 50 miles of air may intervene between the camera and a line of peaks which must have a diamond-clear skyline to shoot successfully, the barely perceptible haziness due to fires 500 miles or more away makes photographic effort quite futile.

The low but steadily rising and thickening bank of murkiness we had noticed beyond the Continental Divide for two or three days was caused, as we learned later from the papers, by forest fires along the Pacific Coast of British Columbia. Harmon was already worrying about it, saying that westerly mountain shots from Castleguard and the Columbia Icefield would be quite out of the question until there was a clearing of the air in that direction. As this desideratum could only be consummated as the consequence of a general rain or snow, there was less threat than promise to us in the gathering storm. We would have preferred to have its breaking postponed until Castleguard was reached and camp made, but were not going to complain about anything calculated to dampen down the rising smoke menace however soon or in whatever form it came.

With a report current that there was only one set of tepee poles available from the snow- and wind-stunted timber of lofty Castleguard Valley, there arose a good-natured rivalry between our own and the Potts outfit as to which should be on the ground first to take possession. Since it was practically out of the question for one pack train to pass another on the way, the prize appeared likely to fall to the party making the earlier start.

The rival wranglers were out before daybreak, searching for horses which had scattered far and wide through the timber in their hunt for grass. For a while Potts' men had all the best of the luck; in fact, his outfit was but two horses short before Baptie had brought down a scant half dozen of our own. But it was that far-strayed pair which decided the issue in the end. They were still missing when all 16 of "Soapy's" had been packed and were ready to take the trail.

Following a night in really good grass, the horses were stronger than at any time since pushing on from the forks of the Saskatchewan. This was a fortunate circumstance in view of the fact that our climb to the level of the Columbia Icefield—all of which we were endeavouring to make in one stage—promised to be an arduous one. Few things are more trying than having to force on a pack horse that—in "Soapy's" picturesque parlance—has no "fuel under its boilers."

A mile and half over a densely timbered ridge brought us down to the main or north fork of the Alexandra, usually called the Castleguard. Tumbling through a half-canyoned valley, its slope was such that gravel and mud had been carried down to the flats we had traversed the previous day, leaving only a channel choked with boulders, many of which were in a state of very unstable equilibrium.

Torrential water swirling over boulders makes for fording conditions almost if not quite prohibitive. A horse can break his leg in the wink of an eye in such a place, while footing once lost may be quite impossible to regain. We took every possible precaution in the two crossings we had to make and felt ourselves very fortunate that nothing worse than wet packs resulted. Several of the horses were carried down a hundred feet or more at both fords, but luckily found sloping bars upon which to clamber out. These, with two or three crossings of the Sunwapta and Athabasca on the Arctic side of the divide, were the most dangerous fords we had.

We were now practically at the fountainhead of one of the main sources of the Saskatchewan, and in a region peculiarly favourable to showing how the great volume of flow we had remarked below gathered in so comparatively short a distance. We were under the very drip of the eaves of the continental rooftree. To the east was the rocky summit of Mount Saskatchewan. To the south the peaks of the Lyell massif glittered in solid, unbroken white. To the west, almost directly above us, Alexandra, Spring-Rice and Bryce towered half way to the zenith. To the north were Athabasca, the Twins and Columbia, but cut off from our vision at the moment by the more imminent loom of the southern bulwarks of the great Columbia Icefield. Most of these peaks were over 11,000 feet in height, two of them over 12,000.

We had penetrated to the very heart of a kingdom of ice and snow, and on a day when all of it seemed to be melting. The glaciers spewed out raging torrents of savage power, and not the last, least finger of shadowed snow but gave forth its trickle of down-streaming water to swell the river in the valley. And besides the stream from melting ice and snow, both mountainsides were streaked with rivulets from countless underground springs.

Many of these subterranean flows were of great volume, notably one which fed a splendid waterfall immediately below where we left the Alexandra for the steep climb to Castleguard Valley. The stream from this fine cataract furnished more than half of the flow of the whole river, yet our subsequent explorations revealed that all of it gushed out of the mountainside not more than 200 feet above the brink of the fall. I am inclined to think that 1,500 second-feet would be an under- rather than overestimate of the flow at the time of our visit. In my own experience, I can recall no spring to compare with it in volume, unless it be the one in the Anti-Lebanon Mountains which feeds the river flowing through Damascus, and which tradition claims sprang originally from a footprint of Abraham.

The trail from the river to the Castleguard Valley was blazed by the Interprovincial Boundary surveyors and has been used several times since by mountain-climbing parties. Steep, slippery and deadfall-choked though it was, the going was infinitely preferable to the punishing grind in mud and water we had had since leaving the Saskatchewan.

And the valley itself—a thousand acres of mountain meadow surrounded on three sides by perpetual ice—was a near paradise.

Although at the verge of timberline, stunted but close-growing fir and spruce provided wood and shelter from the wind. A streamlet flowed past the cook tent door, and knee-deep grass up the valley promised a feast for our half-starved horses that would hold them from all desire to stray.

Nothing but clouds and the smoke of our own campfire could cut us off from the pinnacles of Spring-Rice and Bryce, notching the skyline to the west, or the buttressed heights of Castleguard, lone sentinel of the Columbia Icefield to the north. A 20-foot cascade on the river above the camp, and a 100-foot sheer fall just below, were the crowning touches in a picture etched deep on the tablets of memory.

We came across several deserted camps in Castleguard Valley while scouting a site for our own, and in one of these—doubtless the last to be occupied—was the set of tepee poles for which we had raced. After a single appraising glance at the pile of stubby sticks, "Soapy" announced that, since Bill Potts' outfit would probably come straggling in along toward dark after a late start, it would be only an act of common decency to leave a set of tepee poles all ready for them.

I was consumed with admiration for the brotherly act of the old woodsman, especially since it cost him and Baptie nearly a whole half-day of chopping and trimming to reduce a score of stunted, scraggly fir-trunks—the smallest of them not less than a foot in diameter at the base—to proper tepee-supporting dimensions.

My admiration grew and warmed as I watched those two self-sacrificing packers toiling down with their burdens all through the snow flurries of the chill afternoon. But it died a sudden and violent death at supper when "Soapy," waxing confidential over coffee, revealed the hidden motive behind its apparent generosity. He was badly in need of two strong pack horses to replace the pair of trail-worn scrubs picked up at Lake Louise, and he wanted to coax Bill Potts into a humour to make the exchange. He reckoned that presentation bunch of tepee poles ought just about to turn the trick. And even if it didn't, why no great harm was done nohow. The geesly sticks were too short for our big tepee by all of five feet!

It was highly reassuring to have a sly old fox like that working with heart and head in one's interest. If you couldn't admire old "Soapy" for one thing it was always very easy to admire him for another.

There were three inches of snow on the ground by the time the tents were pitched and before dark a roaring storm was swooping down upon the valley from the Columbia Icefield. Although sheltered on all sides by close-growing timber, the tepee had still to withstand some terrific onslaughts from the swirling wind squalls. The almost cone-shaped poles—many of them weighing over a hundred pounds apiece—gave incomparable stability, but there was no way of shutting out wind. This was principally due to the fact that the savage gusts, caroming off now one side of the valley and now the other, never attacked twice from the same direction. That made it impossible to trim the flies so as to exclude more than one gust out of ten. And so the smoke from the fire—with sparks and occasionally very sizable fragments of glowing embers—were blown all over the inside of the tepee.

Harmon, in his snug little "pup" tent, reported a very comfortable night; the four of us in the tepee were far from happy. "Soapy" and Ulus complained of twinges of ancient rheumatism, reawakened by the wettings of the last few days and the chilling of wind-fanned backs. Rob's bed was set afire from sparks; also my own, the latter in dangerous proximity to the inflatable rubber mattress of my sleeping bag. As a crowning touch, "Buster" and "Tip," fighting over a coat to sleep upon, rolled into the fire and burned their feet, incidentally filling the tepee with the odour of singed hair for the rest of the night.

Six inches of snow on the ground in the morning, with icicles a foot and more long at the waterfalls, were renewed reminders of the shadow line ever dividing summer from winter in the Canadian Rockies, even before the month of August is torn off the calendar.

But the storm had functioned with a restraint beyond all praise. It had spilled down just enough rain and snow effectually to put the smoke to sleep for several days, and then had stopped. The mountains were still too much obscured by clouds for Scenic photography, but for waterfall and river shots the light was all that could be desired.

Leaving "Soapy" to overhaul and patch pack gear, Harmon, La Casse, Baptie and I set off to follow Castleguard Creek to the lower valley. Our maps were a bit obscure as to just how this affluent found its way to the river we had followed up from the Alexandra. We were especially anxious to know if it was the source of the water forming the superb fall at the point where we had started to climb the mountain. And if Castleguard Creek did not feed this fall, it was highly important for us to find out where it *did* come from. Topography was such that there simply *had* to be some striking, if not quite unique, movie shots where the water for that fine cataract tumbled down or gushed out of the mountainside.

Castleguard Creek, after meandering peacefully down the open valley for a couple of miles, began a sharp series of descents at a spreading cascade, 20 feet in height, not far above our camp. Two hundred yards below there was an abrupt fall of a hundred feet, followed by a broken series of cascades which carried the stream down a gorge in what was an almost abrupt limestone cliff.

The main fall above had possibilities for movie shots of enough promise to make it worthwhile waiting for a better lighting of the mountain backgrounds.

We had a difficult scramble down the spray-wet gorge through the cliff but found the several splendid tumbles of water too much broken up for pictures.

When the torrent turned off to the southward in a way that proved beyond doubt it was not a feeder of the great cataract below, we started working along the mountainside to the north. Since the trail we had followed with the horses had crossed no large stream, it seemed reasonable to believe that the one we sought must come down within a very short distance of the point we had now reached.

Stopped by an impassable gorge with a bare trickle of water in the bottom of it, we headed straight down the mountainside, hoping to come to the lower valley somewhere in the vicinity of the mysterious cataract.

Several deer scared up in the woods were left unmolested, the season not yet being open. A flock of spruce fowl, however, offered fair game. Harmon picked off three or four of these most utterly fearless of all wild birds with his "22" pistol. Baptie, with a real reputation as a hunter to sustain, disgraced himself by missing six shots with his "38 automatic" at less than ten feet. Fortunately, the obliging "fool hen," with plumage no more than ruffled by the wind of that salvo of flying "soft-noses," remained on its perch long enough for Ulus to sneak up from the rear and complete our mess of pottage by bringing it down with a stick.

Shooting lapses—as well as birds—of that kind are hard to account for. It was the keen-eyed, agile Baptie who kept us in fresh meat all of the trip, both his marksmanship and strategy in bringing down goat and sheep proving of the highest order. But he was never quite able to live down that altogether inexplicable slip-up of missing six shots at a bird that was almost roosting on the barrel of his pistol.

From that day on the very croaking of a "fool hen" among the distant trees would start a telltale blush surging up Rob's shame-bowed neck.

The roar of the big cataract lured us northward again as we descended, but only to meet another impassable obstacle in a sheer-walled crack in the limestone. When we finally came down to the valley floor, it was to find ourselves just below the big fall, with its full discharge blocking the way to the gravel bar from which our return trail started.

Fording this torrent was too much for one man alone, but we finally decided it could be effected by clasping hands and forming a human chain. This promised to leave at least one unit always with a comparatively solid footing, and so able to brace and steady the others through the deeper part of the channel. I had seen the thing done in Alaska some years previously.

This precarious crossing, with the fall as a background, looked like such good movie stuff that Harmon decided to set up his camera and make a shot. We were bound to win whatever happened, he reasoned. If the chain went over unbroken, it was a good piece of fording; if it broke, on the other hand, and one or two got rolled, it was a good movie. What more could a man ask than a chance to play a sure thing game like that? Thus Harmon.

He had not exactly what one would call a scintillant wit, but was nevertheless often ready with quaint little quips like that which went far toward silvering dark dabs of the clouds overhanging our many leaden days.

Unfortunately for the movie, my human chain functioned exactly as planned. There was one inspired moment when I, as anchor man, was the only link firmly planted on the bottom. A relaxing of my wet fingers, a twitch of the wrist, and the sequel would have furnished a shot as "surefire" on the screen as a facial massage with a custard pie. Nor did I miss the possibilities of the thing at what would have been the psychological moment. Not I. The possibilities were far too obvious to miss; besides, Ulus' strained fingers were all but slipping from mine as it was.

But—well, Rob's temper was still rough and raw over the "fool hen" fiasco, and it didn't seem quite prudent to add what he might have construed as insult to so deep an injury. The tempers of packers, pack horses and prima donnas cannot be pinned down and charted like those of other beings.

And then there was the circumstance of Ulus being the cook of the expedition. There is an ancient law of the Medes and Persians which decrees that a man responsible for the killing or disabling of a camp cook shall take on the job himself. It was altogether improbable that a man would drown in the tumbling cascade below, but, with the stream composed of about equal parts of water, sand and rolled boulders, I am inclined to think that Lloyds would have quoted a high premium on a disablement policy.

All in all, then, it is probably just as well that I went ahead and enacted my part of human anchor according to the original plan. But what a movie we missed!

Between the cataract, the swirling chute and the swaying of the human chain, the shot was full of lively action even as it was. When it was over we recrossed to ferry Harmon and the cameras. On the third crossing, with the movie no longer set up to record events, we missed the big spill by an eyelash. The near upset was caused by "Buster," who, plunging in above, was carried down to lodge against Harmon's legs.

With the photographer's squat figure already in hair-poised balance as the fierce midstream current beat against it, a chip would have been enough to start it toppling. "Buster's" floundering anatomy was more than a chip, and so Harmon lost his hold with both feet. That yanked down Baptie and half upset my own balance. The sturdy La Casse, fortunately, was in shallower water and well braced. His firm hold gave me a chance to recover my footing, after which it was not difficult to steady the uttering of the outer end of the chain. A complete mess up at this moment would have been a serious one, as the loss of at least the motion picture was inevitable.

"Buster" was swept down 200 feet—much of the distance

underwater—before pawing against a bar along which his stout legs carried him out. He was coughing water out of his lungs when he rejoined us, and acted rather after the manner of a dog which suspects that a trick has been played upon it. Half a minute later "Tip" breezed down the bar from above with a dry back and not a hair ruffled above the toes with which he had been spurning the thin layer of snow still lying in the timber. He seemed mightily pleased with himself over something, probably the fact that he had completely avoided a crossing so rough as to have had even his strong-swimming mate in serious difficulties.

The significance of "Tip's" dry hide suddenly broke upon us. It meant that, unless he had crossed on a fallen tree or a natural bridge, the whole stream must come gushing out of the mountain within a very short distance of the brink of the fall. And that, as I have already told, was what a three- or four-hundred-foot climb revealed. Just what sort of a sixth sense apprised "Tip" of the way things were disposed remains as much of a puzzle to me as did the mainspring behind his action at the ford of the Alexandra.

The stream was almost certainly an underground drainage from a part of the Columbia Icefield. From the fact that it appeared to vary but slightly in volume while the glacial streams on the surface were rising and falling greatly under the influence of the weather, it would seem probable that it came from a reservoir of great capacity. Subsequently we found many sinks in upper Castleguard Valley where the waters of melting snows disappeared into the earth. It is conceivable that much of the perennial glacial meltage goes underground in similar fashion.

On the way back to camp La Casse led us to a large limestone cave which had been discovered by a mountain-climbing party some years previously. It extended back several hundred feet into the mountain, while from its mouth ran a channel which gave evidences—in whorls of sand and back-laid blades of grass—of having recently carried a torrential flow of water.

La Casse described how, when camped in Castleguard Valley

with the Thorington party in June of the previous year, they had been attracted to the cave by a tremendous roaring. This thunderous sound broke forth without warning every afternoon at almost exactly four o'clock, continuing far into the night. The following morning the channel was dry, remaining so until the flood came down again in the afternoon. Their theory was that, like the flow from the cave on the brink of the wall of the lower valley, this one was fed from the drainage of sinks under or near the main icefield.

This upper cave was only a few hundred yards from our camp, which gave good opportunity to keep it under close observation. We were especially anxious to picture the Jove-like birth of the underground river; also to make a series of shots of its waters tumbling down a titanic limestone stairway immediately below.

These giant steps, as regularly and evenly set as the sides of the Cheops Pyramid, and at about the same general slope, were formed by the breaking back of the lofty cliff we had clambered down in following Castleguard Creek. It is probable, indeed, that the gorge cut by this intermittently flowing river was the identical one which had turned us back in our search for the source of the great cataract below.

In spite of repeated visits to the cave during the several days we remained in Castleguard Valley, we were never able to find it in flow. This was a great disappointment, as the huge discharge of water tumbling down those several hundred feet of evenly spaced stairs would have made a picture both unique and beautiful.

The irregularity of flow of the river from this upper cave would seem to indicate that its waters had their source in snow meltage rather than in that from the glaciers. This would account for the fact that it was running in June—a time when the early summer thaw is at its height—and not in August, when most of the general snowfall is gone.

Returning to camp in the early afternoon, I found my first opportunity to disentangle, clean and dry the pulpy bundle of wires, batteries, boxes and packing that was once a radio outfit. The cedar pack box was badly split and dented from the terrific blows sustained

Lewis Freeman, Buster the dog and radio (WHYTE MUSEUM OF THE CANADIAN ROCKIES, V263/NA-2250)

Byron Harmon with camera at Fortress Lake
(PHOTO BY LEWIS FREEMAN. WHYTE MUSEUM OF THE CANADIAN ROCKIES, V263/NA-2400)

Rob Baptie and Lewis Freeman
(WHYTE MUSEUM OF THE CANADIAN ROCKIES, V263/NA-2248)

Castleguard Falls
(WHYTE MUSEUM OF THE CANADIAN ROCKIES, V263/NA-2301)

Lewis Freeman working on the radio at Bow Lake camp
(WHYTE MUSEUM OF THE CANADIAN ROCKIES, V263/NA-2277)

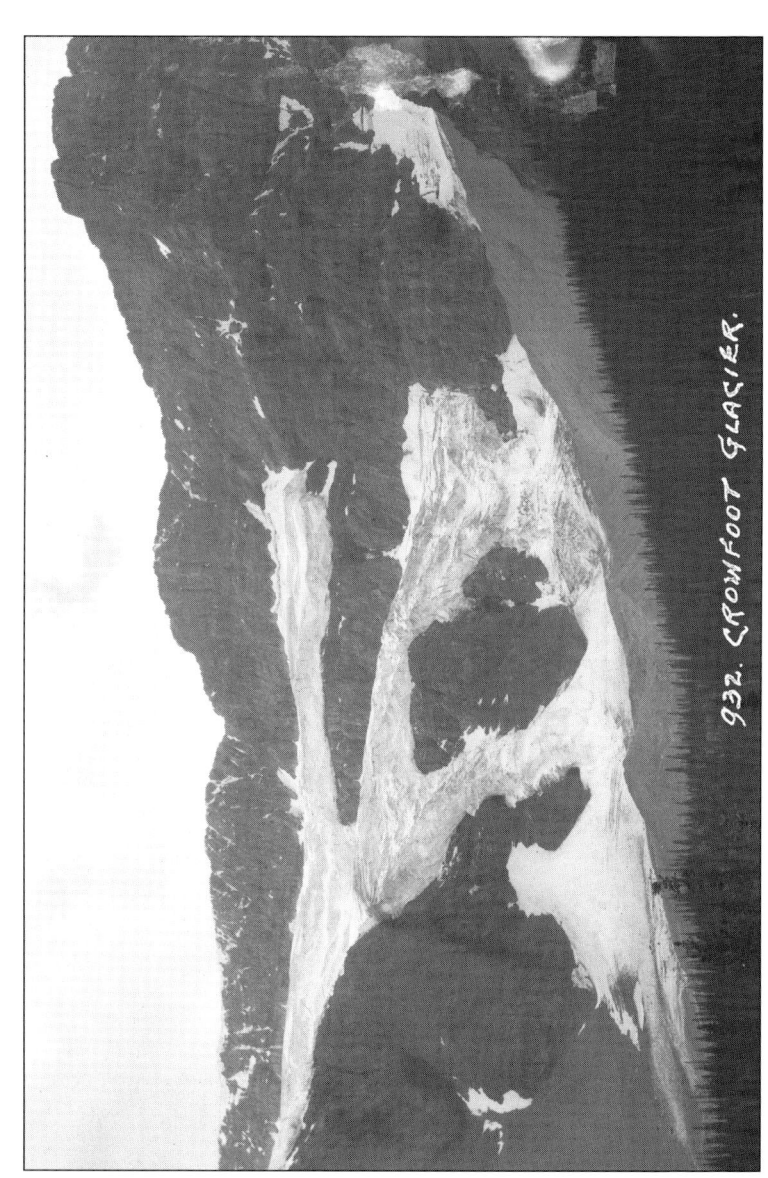

Crowfoot Glacier at Bow Lake
(WHYTE MUSEUM OF THE CANADIAN ROCKIES, V263/NA-5727)

Byron Harmon working on the radio
(PHOTO BY LEWIS FREEMAN, WHYTE MUSEUM OF THE CANADIAN ROCKIES, V263/NA-2249)

Mount Lyell, Castleguard Valley
(WHYTE MUSEUM OF THE CANADIAN ROCKIES, V263/NA-2143)

Fording the Alexandra River at Castleguard Falls
(WHYTE MUSEUM OF THE CANADIAN ROCKIES, V263/NA-2425)

Party of four on Castleguard
(WHYTE MUSEUM OF THE CANADIAN ROCKIES, V263/NA-2161)

Saskatchewan Glacier
(WHYTE MUSEUM OF THE CANADIAN ROCKIES, V263/NA-2284)

Top of Sunwapta Falls
(WHYTE MUSEUM OF THE CANADIAN ROCKIES, V263/NA-2412)

Mount Columbia
(WHYTE MUSEUM OF THE CANADIAN ROCKIES, V263/NA-2264)

Maligne Lake campsite
(WHYTE MUSEUM OF THE CANADIAN ROCKIES, V263/NA-1328)

Pack train on the Poboktan
(WHYTE MUSEUM OF THE CANADIAN ROCKIES, V263/NA-2422)

Mount Coleman from Pinto Lake
(WHYTE MUSEUM OF THE CANADIAN ROCKIES, V263/NA-2388)

from rocks and trees, but was still held together by its close-set brass screws and a double lashing of fishing line. Its contents had gained rather than lost in volume since we had last tied it up on the trail from Bow Pass down to the Mistaya. Most of the accretion was due to the all permeative glacial silt which, besides depositing itself in a half-inch-deep layer at the bottom of the box, had smeared everything inside with thin, clinging blue mud.

The little black box o' tricks itself, with not a right angle left, gave forth a noise like a child's rattle from the screws, switches and various odds and ends of breakage shaking about inside. Once it was trued up and the knick-knacks restored to place, however, it was just about the same as ever, at least so far as looks were concerned. And looks were, of course, the only thing that mattered much now. We had long since given up hopes of using the bruised and battered bundle of junk as a radio receiving set.

The tubes, also (there were eight of them including four spares), appeared to be reduced to a condition in which they were better adaptable to movie than to radio use. While the outer glass of every one of them, protected by soft cotton wrappings in a fibre case, was unshattered, a telltale tinkle from the inside seemed to indicate something radically awry.

It was just as Harmon was proposing setting up the tubes in a row for pistol practice that La Casse discovered a book of directions in the mud at the bottom of the pack box. After restoring the first page to legibility by dabbing it in his dishwater, Ulus puzzled over it for a while before interposing with a request that we defer our target practice long enough to give him a fair chance to bring the radio back to life. He had discovered a diagram showing the proper hookup of the box and batteries, he said, and we already had special directions for stringing the aerial. If by any chance the tubes and the batteries still retained their life, who could say that we should not be bringing the outside world to our tepee door inside of an hour?

As it is always good policy to humour a camp cook, we let the optimistic Ulus have his way. Genius that is denied opportunity for outward bloom often strikes inward. With two months of hard work still ahead it was not well to have anything rankling in the mind of any member of the party, least of all in that of the cook.

The batteries spat forth strong, vital sparks when we began connecting them up, but the real surprise came when each of the tubes, tried out one after the other, responded with a bright red glow to the circulation of the current. There was no possible question of their being very much alive and ready to play their part of the game. Just what was the loose hardware clinking around inside of them we never did learn. Probably it was only a fragment of glass from the collar at the base of the filament. Certainly, in any event, it was a part of no essential connection.

The aerial, by dint of rough climbing and some very clever lasso-like casting of the loops of the insulated wire, was finally strung between a ledge of rocks and the last scraggly pine at the edge of timberline. A jiggering of the dials and switches along toward the end of the afternoon seemed to reveal restless stirrings of the ether, but not all our cunning could resolve it into anything more coherent than clangs and howls.

Further study of the book seemed to point to the various units of our wiring system as the probable seat of trouble. The receiving set had been connected to the north rather than to the south end of the antenna, while a loose double bow-knot of insulated wire could hardly have been expected to give as satisfactory an outlet for the lead as the soldered joint recommended by the book.

All the noises of a busy, booming world seemed to be lying in wait to swoop down upon our quiet alpine valley when reconditioning was complete and we put on the headphones and started worrying the dials again. With many stations appearing to be crowding, like a line at a ticket window, waiting to be the first to get to us, and with scant cunning of hand or ear in the matter of tuning in or tuning out, it was slow work reducing the chaos to a semblance of order and making the clamouring voices take their turn.

The long northern twilight was over and darkness descending

upon the ice-bound valley before a plaintive ballad, in a voice like the wail of a soprano coyote, was heard announced as being broadcast by an Oklahoma City station "for the benefit of the auto-campers." The song was "Carry me Back to Ol' Virginny," which seemed a bit inappropriate considering the fact that most of the auto-campers must have come from west of the Mississippi.

The next turn of the dial brought an unpleasant shock. I tuned into a thunderous roar, like the clamouring of hungry lions at feeding time or the howling of the stage-trained Roman mob in the pauses of Marc Antony's oration over the body of Cæsar. Then there was silence, quickly cut into by a strident voice shouting, "The great Amerikun peepul will never stand for being thus strangled by a tyrant!"

It was the open season for politicians, of course, and this must have been one of the La Follette minions calling to his mates, or, rather, his dupes. No South Sea schooner skipper avoiding an imminent coral reef ever threw his wheel harder over than did I the radio dial in my haste to put a thousand miles of pure ether between our uncontaminated valley and that howling demagogue.

We kept tuning in on these pestiferous harangues all the rest of the trip—but they were gone at the twist of a thumb. It is no end of a pity that the primal founts themselves cannot be stilled in the same manner. A twist of the thumb, properly applied, really goes quite a ways toward quieting the bobbings of a wind-fanned Adam's apple.

Following directions still more carefully in restringing the aerial the next day, we were rewarded by picking up Calgary, with a score of stations scattered between Los Angeles and Pittsburgh, shortly after it was dark at 8:30. This highly interesting concert was brought to a sudden end, in the middle of a tenor solo broadcast from CKCD, Vancouver, when a brief but violent squall from the icefields broke off short the limb of the dead tree to which the antenna was attached.

The next day we changed the set-up so that the lead ran down into the tepee and we could listen-in from the comparative comfort of seats on our bedrolls and with the light and warmth of a crackling pitch-pine fire. And here it was, for the final two months of the expedition in the Rockies, that the world came to visit us almost every night through the medium of the little black box which we had come so near to throwing into complete discard.

Without giving the lengthy list of stations which we had at one time or another, I may say that these included a majority of the high-power ones east of the Mississippi, others as far south as Baton Rouge, high- and medium-power stations in California, and practically every station in the Pacific Northwest, British Columbia and Alberta.

The highlights of the radio adventure I will touch upon as they developed. From Castleguard on it was one of the most enchanting features of the trip.

Chapter VII

THE MOTHER OF RIVERS

THE MAIN PIECE OF WORK planned to be done from our camp at the edge of the Columbia Icefield was the filming of the panorama from the summit of Castleguard. This striking peak, although but little over 10,000 feet in height, is so located as to offer a vantage quite unique for viewing or photographing the great Columbia *mer de glace*, with the stupendous walls of lofty mountains surrounding it.

We were anxious for at least two hours on the summit during which there would be not only local sunshine, but a clear skyline of peaks in every direction. Such an ideal day might occur not over once or twice in a summer, but we were prepared to wait for the nearest possible approximation to it.

The sudden breakup of the humid Chinook spell, followed by a short but violent snowstorm of almost blizzard magnitude the day of our arrival, seemed to point to a long, tedious wait. Then another kaleidoscopic shift brought just the day we were looking for, or at least a morning which promised to develop into such a day.

As practically the whole 3,000 feet of climbing was over ice or snow, there was no chance to use the horses.

For a real mountain-climbing party the ascent of Castleguard is comparatively easy. Harmon and La Casse were fairly experienced climbers, but the packers and I were novices. Added to this, was the fact that we divided between the five of us—what with the heavy movie camera and tripod, four still cameras and much incidental impedimenta—an average of possibly two or three times the weight ordinarily carried by the alpinist.

And so we made a hard tiring pull of it, with many halts interspersed to regain breath and to massage incipient cramps out of knotting muscles. The last thousand feet was the steepest, steps having to be cut with an ice axe for most of the distance. Personally, I was extremely glad of the sustaining pressure of a rope during the precarious clamber up a "chimney" just below the summit.

Possibly lacking the sheer breathtaking wonder of the first sight of Kinchinjunga's snows from Darjeeling, the view from the summit of Castleguard is still one of the great mountain panoramas of the world. Set on the southern rim of the Columbia Icefield, with no other peak encroaching on its domain for many miles, there are no masking barriers close at hand to cut off the view in any direction.

Not only are almost all of the great peaks of the Canadian Rockies system notched into the splendid panorama, but also many of those of the Selkirks and the Gold Range, far beyond the purple-shadowed depths that mark the great gorge of the Columbia River.

The first white man to see the Columbia Icefield was J. Norman Collie, in 1898, who wrote the following interesting description of the panorama unfolding from the summit of Mt. Athabasca:

> A new world was spread at our feet; to the westward stretched a vast icefield probably never before seen by human eye, and surrounded by entirely unknown, unnamed and unclimbed peaks. From its vast expanse of snow the Saskatchewan Glacier takes its rise, and it also supplies the head waters of the Athabasca; while far away to the west, bending over in those unknown valleys glowing with the evening light, the level snows stretched to finally melt and flow down more than one channel into the Columbia River, and thence to the Pacific Ocean. Beyond the Saskatchewan Glacier to the southeast a huge peak (which we have named Saskatchewan) lay between this glacier and the west branch of the North-Fork, flat-topped and covered with snow, on its eastern face a precipitous wall of rock. Mt. Lyell and Mt.

Forbes could be seen far off in the haze. But it was to the west and northwest that the chief interest lay. From this great snow field rose solemnly, like 'lonely sea-stacks in mid-ocean,' two magnificent peaks which we imagined to be 13,000 to 14,000 feet high, keeping guard over those unknown western fields of ice. One of these, which reminded us of the Finsteraarhorn, we have ventured to name after the Right Hon. James Bryce, the then President of the Alpine Club. A little to the north of this peak, and directly westward of the peak Athabasca, rose probably the highest summit in this region of the Rocky Mountains. Chisel-shaped at the head, covered with glaciers and snow, it stood alone, and I at once recognized the great peak I was in search of; moreover, a short distance to the northeast of this mountain, another, almost as high, also flat-topped, but ringed around with sheer precipices, reared its head into the sky above all its fellows.... At once I concluded that these might be the two lost mountains, Brown and Hooker.

Bryce is still the dominating peak in the panorama unfolding from most of the Columbia Icefield, but its ascertained height of 11,507 feet falls considerably short of Collie's estimate. The two high northern peaks, supposed to be Brown and Hooker, were those since named respectively Columbia and Alberta.

Columbia, with an altitude of 12,294 feet, is indeed the highest mountain in this region of the Rockies, and ranks second only to Robson in the whole system. Seen from anywhere to the south, however, it appears a comparatively insignificant peak, probably due to the foreshortening of the long slope up from the great icefield. This rounded hummock of snow can hardly be recognized as the slender Matterhorn-like pinnacle we were later at such trouble to photograph from the head of the Athabasca.

Presenting even more contrasted views as seen from the south and north is the Snow Dome. Although the altitude of this remarkable eminence is 11,340 feet, the slope to its summit from the Columbia Icefield looks gradual and smooth enough to be navigated by an automobile. Yet the northeast side of the mountain is an almost sheer cliff of 5,000 feet to the upper flats of the Sunwapta.

It is the Snow Dome which forms the hydro-graphic apex of the drainage of the Columbia, Athabasca and Saskatchewan.

The deep purple-indigo dome of the sky looked clear enough to last a week as we struggled up to the top of Castleguard, but Harmon was too sapient of Rocky Mountain summer weather to presume upon momentary fairness. Casting his eye rather to where a vaporous boil of dark clouds was beginning to surge against the barriers above the gorge of the Columbia, he set up his cameras and began cranking off film at feverish haste.

Two complete panoramas were made, as described in the chapter on scenic work. One was shot with, one without, a filter on the lens. The panoramas were followed by shots with the telephoto at individual peaks, and "slow-turn" exposures on the progress of the gathering clouds across the face of Bryce. A round of shots with the various still cameras gave us everything we came for, and under almost perfect conditions of light and air.

We finished in nick of time. Before the cameras were packed up the vanguard of the mists were pushing up the slopes from the west, to come swarming over the top as the last man took his place on the rope and we started to descend.

Going down, as is usually the case with a novice, I found rather more trying than climbing up. There were two or three times, when my dangling heels refused to find holds in the "chimney," that the rope was more than an ornament.

Hardly had we kicked free the rope as the foot of the steepest pitch was reached, than La Casse loping on ahead with the dogs over apparently unbroken snow, dropped completely out of sight. He reappeared an instant later, having caught one foot on a ledge just under the brink of the crevasse.

The latter, thinly but completely bridged by the drift from the late snowfall, was of great depth, smooth green walls disappearing from sight far below the length of our rope.

In the interesting record published by Dr. J. Monroe Thorington, F.R.G.S., of his ascents of a number of mountains of the Columbia Icefield in 1923, I find this sentence concluding the account of the return from Castleguard:

> It was a day of enjoyment for all, although the disappearance of our cook in a small crevasse frightened us badly.

By an interesting coincidence, it was not only the same cook who broke through into a crevasse, but almost certainly the same crevasse that was fallen into. As old "Soapy" commented: "Seems like there was two things cooks was most always pretty near losing—their tempers and their hides."

As a matter of fact, La Casse was not only extremely good-tempered—away from the cook-pots and the smoke of bacon grease—but also probably the best mountain man in the party.

Practically all of the important peaks surrounding the Columbia Icefield have been climbed, though two or three of the highest of them were only conquered during the summers of 1923 and 1924. The most notable climbing work in this region was undoubtedly done by the Thorington expedition in the former year. Besides several minor climbs and traverses, first ascents were made of the North Twin and Saskatchewan, a second ascent of Columbia and a third ascent of Athabasca.

The North Twin is 12,085 feet in height, the only loftier peaks in the Canadian Rockies being Robson and Columbia. Thorington's account of the initial ascent is a classic of modesty and brevity.

"It is a simple story," he writes in the *Alpine Journal*. "We saw our peak, walked toward it, up it, and back again. There was only the distance."

That comprehensive second sentence has almost the succinctness of Cæsar's famous "*Veni, vidi, vici,*" without the latter's offensive bombast.

A hint of the effort involved in this fine ascent is revealed in the concluding paragraph of Dr. Thorington's notes,

"No one who does not follow in our tracks will quite understand that journey back across the endless icefield. The exhausting first half-hour in a little blizzard, obscuring the trail 20 feet ahead; clearing, with a crimson, gold, and orange sunset banded against lead-blue storm-clouds behind The Twins; the unearthly light in the snow-banners and mist above Columbia; the soft, rosy haze filtering into the distant Selkirks, lifting them up and making them unreal.

"We were too tired to appreciate it, plodding on and on, in deep, insufficiently crusted snow, over plateau and ridge and dip, until darkness came. The field is so huge. In one corner the stars were out; in another, beyond Mt. Athabasca, dark clouds hung and lightning flashed. We lit our lantern and went on through the night, pulling into camp at last, with morning light upon the hills as it had been 23 hours before when we departed."[3]

The two days following our ascent of Castleguard were spent in preparing for what seemed likely to prove the roughest traverse of the trip—the crossing of the long easterly tentacle of the Columbia Icefield known as the Saskatchewan Glacier. Preparations consisted not only of putting pack gear and the general outfit in as good a condition as possible, but also of a day of advance exploration on the glacier.

With supplies materially reduced by both consumption and attrition, the packs were going to be much lighter than at the outset. This, and the fact that the condition of the horses was improved greatly by a week of rest and grazing in the mountain meadows, was a point in our favour. Heartening, also, was the certainty that there was to be no timber to knock off the packs and no mud to bog down the horses. Our worst troubles would come from the ice and rocks, and of these there was not much that could be learned in advance.

The protection of the providentially restored radio was now one of our chief concerns. Fearing further serious smashings, or possibly total loss, if a pack horse stumbled into a crevasse on the glacial traverse, La Casse spent his spare hours fashioning a rude sledge on which the precious box was to be lashed and drawn by man- or dog-power across the stretches most dangerous to the horses. The plan seemed perfect as we discussed it in theory; in practice it developed serious defects. Fortunately, we were able to give it a preliminary test.

The sledge proved easy enough for a man to draw over ice or snow on the level, or even up a considerable slope. On a side-pitch, however, one or two extra men were necessary to keep the top-heavy load from upsetting. As it seemed certain that all hands would be needed to manage the horses once we were out on the glacier, a full sled-crew was out of the question.

"Buster," with a couple of seasons' experience in the Banff dog races, snaked the loaded sledge across the ice without effort. A man or two at the side to prevent upsets still proved necessary, however, and with no extra men likely to be available we were forced reluctantly to abandon the whole idea of sledge transport for the radio.

Unluckily, we neglected to unharness "Buster" when the trials were over. That was how both sledge and radio formed a comet-like wake astern of a flying husky that started off full tilt at the challenging chirrup of a rock-rabbit. In spite of the hundred-pound drag, "Buster's" tigerish spring actually carried him all the way across the small crevasse suddenly intervening between him and his quarry. Sledge and radio, of course, plumped right into the hole in the ice, taking with them a very surprised and frightened dog. By good fortune it was a very shallow crevasse, and so we were able to fish "Buster" and his burden out of the bottle-green depths with the aid of an alpenstock.

Our purpose in traversing the length of the Saskatchewan Glacier with the pack train was twofold. On the one hand, that route would save from 40 to 50 miles of flooded flats in getting around to the north side of the Columbia Icefield; on the other, it would give us a far more intimate glimpse of the great icefield itself.

Heavy as the possible penalty of failure might be, the chance was deemed worth taking on either score. We were the more sanguine of success from the fact that this same route had been traversed the previous year by the Thorington party.

La Casse, who had been along on this initial venture, which was undertaken in June, thought that we would find conditions less difficult in September. This was on the perfectly sound theory that, with most of the winter's snow gone at the later date, there would be firmer footing on the solid ice of the glacier, with less chance of stumbling into thinly covered crevasses. Like the radio sled, this was good theory but failed of vindication in practice.

Just as the most serious problem of a man jumping out of a balloon is that of landing, so that of the glacial traverse was getting off over the lower end. La Casse had warned us of this, and a day's advance scouting on foot had proved the warning in good point. It was going to be rough getting onto the glacier, and still rougher travelling across and down it; but the very worst that seemed likely to happen at the snout was that the horses might have to be thrown and let down on their sides by ropes.

These observations, with the deductions drawn therefrom, were all right as far as they went. The trouble was that the preliminary scoutings were not carried quite far enough to reveal that rockslides had filled or carried away a short but very important bit of mountainside traversed by the previous party in reaching the flats below the glacier. This we were not to learn until the bridges had been burned behind us by sliding the horses down a slope of ice up which they could not be taken back. What followed is not pleasant even to write about four months later.

Anticipating difficulties on the ice, we had broken camp and started off up the valley at a very early hour on the morning of September 6. An hour and a half of steady climbing took us over the divide at an elevation of about a thousand feet above timberline. A mile almost on a level, with a slight descent at the end, brought us out by the side of a beautiful little ice-walled lake at the side of the glacier.

What appeared to be the easiest point to reach the ice—a rounded depression at the foot of the lake—had been proved impracticable because of a broad swath of unfathomable glacial mud.

The only approach was across the broken rock ridges above the head of the lake, where we had already found evidences of the passage of the previous party. It was terribly rough going here, especially where the ice and rock were mixed in a vile conglomerate; but the pack train was worked across without serious trouble.

At the edge of the ice the horses were tied head-to-tail in bunches of three or four, each unit to be led by a man on foot. This was to give more complete control and especially to prevent straying when out among the open crevasses.

La Casse with one bunch took the lead, Baptie, "Soapy" and I following with the other three. Harmon, leading only the movie horse, scouted on a roving commission after pictures.

The horses, although in fine fettle after five days' feasting among the juicy mountain meadow grasses, took the work seriously and showed little desire to bolt or hang back. Plainly the strangeness of their surroundings and the unstability of their footings had a sobering effect.

The day was even more nearly perfect for photographic work than that on which we had ascended Castleguard. There was a clear vault of indigo sky overhead, with the west banked full of rolling cumulus clouds which gave all the effect of an approaching storm with nothing of the threat.

The rims of the hanging glaciers high up on the mountains to the left and right were scintillant with reflected sunlight while on all sides of us the broken surface of the icefield threw tremulous shadows like those of the waves of a rough choppy sea.

The pictorial possibilities of the traverse so smote upon the artistic spirit of Harmon as to leave him for the moment gasping like a child set down in a candy shop and told to help himself.

Winding and twisting, doubling and turning, La Casse slowly worked the pack train out toward the great lateral moraine which stretched down the middle of the glacier like a half-completed railway embankment.

We had already observed that, while the horizontal going was rougher along the edge of this dragon-back of up-pushed rock, the chances of perpendicular descent into a crevasse was rather less than along the sides of the glacier.

Repeatedly turned back by yawning cracks opening beyond eye scope into the blue-green depths of the ice, there was none long enough to block the way completely. Such chances as there were of breaking through, La Casse took by going ahead.

Indescribably rough and slippery as was the surface of the ice, it proved unexpectedly firm underfoot. A horse had its tail or neck sharply stretched now and then, as its mate immediately astern or ahead went down and wallowed at the end of the binding halters. The casualties to packs, however, were rather less than when working through the mud or in the timber.

The abysmal groans of cracking ice were a bit nerve-racking to the novice at first, but one steadied to them in time when they failed to make good their threats of opening up the glacier under his very feet.

Doing two or three miles of windings for every one of progress toward the foot of the glacier, the hours slipped swiftly by. We were still a mile from the point where the end of the ice wall of the snout broke down to the flats of the Saskatchewan when a deepening and widening of the crevasses ahead warned that it was time to get the pack train off onto the rocky mountainside to the north.

The only place where it appeared this could be accomplished successfully was a narrow tongue of ice and gravel, running down to the native rock between two very deep crevasses. The "runway," with the ice breaking sharply off on either side, was about 20 feet; the distance down to the rock perhaps 80.

With the slope far too smooth and steep to give hoof-hold, the only really safe way of getting down would have been to unpack the horses, throw them, tie their feet, and lower them on their sides by ropes. An animal was bound to go down somehow, of course, once he was started. The danger was that, floundering in his fright, he might get into one of the crevasses.

With not nearly enough time to resort to the safety-first method, nothing remained but to do the best we could with the only other alternative. This consisted of the simple procedure of leading each horse to the edge, starting him down, head-first, by a rope pulled from below, and trusting to instinct to keep his centre of gravity low by sitting on his haunches and sliding rather than trying to flounder to a footing.

This succeeded even beyond our hopes. Though all of the horses did not have the sense to slide, the three or four that invited rolling by trying to gain their feet kept right on going in a straight line, reaching the bottom unhurt.

With a faint trail streaking the rocky mountain wall where the pack train of the previous year's traverse had picked its way it now seemed that the worst was over. Telling ourselves that another hour at the outside would bring us to a camp in the flats below the snout, we took our time in relashing packs for the rough clamber down along the side of the glacier. The mauve shadows of coming night were piling thick in the ice-filled valley by the time we were ready to push on.

A quarter-mile of increasingly difficult going brought the head of the pack train to the point I have mentioned where a breaking away of the mountainside above had choked the way with a slide of broken rock, extending as far ahead as one could see. This seemed so completely out of the question to cross that we turned to the glacier again in the hope that the ice could be followed at least far enough to get below the rocky barrier.

A way to the glacier over the easy slope of a drift of old snow lent momentary encouragement, only to make the disappointment the keener when we found ourselves on a patch of ice, hardly a hundred yards in diameter, which was completely encircled by a connecting series of impassable crevasses.

With no way of any kind open even to retrace our way up the

glacier, there was nothing left to do but return and try to pass the great rock slide.

The problem with which we were faced here was a dual one. Not only were the huge, sharp-edged fragments so set that the sliding hoofs went down between them at every other step, but many of them were poised in so delicate a balance that the least touch was likely to send them bounding on down to start a slide that nothing but the side of the glacier could stop.

It was a place which, personally, I would have been reluctant to venture upon on foot; with a pack train it seemed like courting certain and serious disaster.

For a hectic interval at the outset, indeed, it looked as though disaster was coming to meet us even before we had made a real start. We were working the animals, one at a time, out onto what seemed the least threatening line of passage across the broken rock, when the horse carrying the radio, trying to find a way of his own, stepped between two tilted chunks, each larger than himself.

Instantly both closed down upon him, smothering his frantic flounderings in the grip of a vise of stone. It looked like broken legs and crushed ribs; almost certainly a case beyond all treatment save that of the revolver.

Baptie and La Casse, who were nearest, started at once to work away the inclosing rocks, the rest of us standing by to prevent a stampede of the horses.

"Buster" and "Tip," who had been chasing rock-rabbits farther up the slide, came charging down, dog-like, for the centre of disturbance. The flying feet of each started rocks rolling almost simultaneously. "Buster" scored the first hit when the solid little hunk he had spurned, smiting Baptie's mount on the off hind hock, knocked it sprawling.

"Tip's" salvo, with a ton more weight behind the main unit of it, found nothing but the back of my knees to mar the curve of its trajectory. Flicking my legs up into the air, it passed on to crash against the glacier wall, leaving me sitting upright in its sliding wake of debris.

A missile heavy enough to have shattered a leg of the Colossus of Rhodes had brushed my relaxed anatomy aside without inflicting more than a few very light bruises.

It took a great deal of lifting, hauling and prying to dislodge the rocks from the imprisoned horse without starting another slide. When this was finally accomplished, however, we had the agreeable surprise of finding the sturdy "Wolverine" practically uninjured beyond much abrasion of hide.

The stout cedar pack box containing the radio outfit had taken most of the crushing on one side, while a grub-box had absorbed it on the other. That the radio was a total loss this time seemed beyond all doubt.

Our progress across that slide of tilted rocks was glacial in its slowness, punishing in its severity. Horses were down repeatedly, usually with the loss of a pack that had to be rethrown while the whole train waited and fretted.

But foot by foot we worried along, and for the first and last time in comparative vocal silence. Even voluble old "Soapy" bethought himself to choke back the explosive oath that might have set the cumulative air-wave wriggling that would start another hair-balanced rock on the slide above.

And finally we won through to solid footing again. This was a sloping cliff of torn and riven bedrock left exposed by the melting of the glacier. It slanted downward like the roof of a Gothic steeple, while its rough surface tore the unshod hoofs like a rasp.

But there were no more rock slabs to slip between, no more balanced boulders to dislodge.

There was need of haste now to make the most of the dying daylight, and tongues and riatas were unleashed in unison to push on the scurrying, stumbling mob. Fantastic cowboy oaths clove the clear air like the spatter of shrapnel and the blended sobbing of winded horses rose in a whistling croon like the wind through the bare limbs of the dead pines fringing the imminent timberline.

Over a half-mile of rock, wet with blood from hoofs that had worn to the quick, we clattered down onto the packed white gravel of the flats below the jutting ice-snout just as the reflection of the last of the afterglow was fading from rose-pink to cinnamon and dusky olive on the summit of Saskatchewan.

A half-mile down from the low, uninteresting forefront of the glacier, across flats whose flinty rock-fragments must have been terribly painful to the worn, torn hoofs of the horses, "Soapy" turned off to a gravelly bench at the foot of the northern mountainside. Someone had told him there was a camping place here, but a narrow strip covered with nothing but rocks and burned timber gave scant encouragement.

With no grass in sight, the only thing to do was to keep going until we found it. Starving pack horses already wearied and battered to the point of collapse could hardly have failed to have serious consequences.

Ascending a steep ridge where the way through the burned deadfalls had repeatedly to be cleared for the horses with an axe, we went on down the other side to find ourselves in a little valley draining to the North Fork of the Saskatchewan a mile or two below. Here there was grass but no water, forcing us to continue on in the deepening twilight.

A quarter-mile farther through dense second growth brought us to the trampled mud banks of a sink a couple of hundred yards in diameter. In the spring and early summer there must have been a lake here, fed and drained by the stream draining the valley of the meltage from the snows. Between evaporation and seepage through the porous rocks this had now been reduced to a shallow pool, foul and ill-smelling from the many animals that had waded in it for water.

With no certainty that we would not fare worse by going farther, "Soapy" reluctantly decided to unpack where we were. That was the only camp on the trip at which good water was not available.

Too tired to cut tepee and tent poles in the dark, we rolled to blankets and sleeping bags as soon as a greedily wolfed supper was over. No one but the dogs could have told at what time the big bull moose, whose fresh tracks we discovered in the morning, stalked straight through the camp on his way to water.

Chapter VIII
OVER TO THE ARCTIC BASIN

As "Soapy" was anxious to work the soreness out of the horses by action rather than to let them grow stiff in a region where the grass was so poor, we broke camp early the next morning and started on the hard, steep climb over the shoulder of the Athabasca to the north. The pack train attained the summit of the ridge and began the descent to Nigel Creek at an elevation of about 8,000 feet while Harmon and I climbed a thousand feet higher to take a panorama of the surrounding mountains.

From a lofty vantage previously utilized by the Boundary Survey we had an extremely fine view of the remarkable ridge of rock which is thrown completely across the valley of the North Fork, forming a natural bridge which has been rarely visited and never satisfactorily photographed.

Following the pack train to the valley on the north, we crossed the almost imperceptible divide of Sunwapta Pass, separating the drainage to the Arctic and to the Atlantic, to find camp pitched four miles farther down almost under the snout of the great Athabasca Glacier.

Here we spent four days, resting the horses and climbing over and photographing the glacier, which vies with that of Saskatchewan for the distinction of being the largest tentacle of the great main sea of ice above.

In spite of violent snowstorms roaring up the valley of the Sunwapta from the Arctic, the radio, still in commission after the terrible banging it had received crossing the glacier, picked up stations in all parts of Canada and the United States. We were troubled practically not at all with "fade out," being able to hold the programs of the stronger stations from end to end when desired.

It was at this camp that we first began to take notice of the Pacific Coast weather forecasts, broadcast by the powerful KGO station of Oakland. These bulletins subsequently became of great practical utility to us in planning ahead photographic work and trail movements of a character to be affected by the breaking of a storm working inland from the northwestern Pacific.

Athabasca Glacier was one of the double-starred points of Harmon's itinerary as originally planned. Visiting there ten years previously by working south from Jasper by the valleys of the Athabasca and Sunwapta, he had found the snout of this tumbling ice river culminating in a perpendicular wall of great height. Though unable to take any satisfactory photographs because of bad weather, he had always recalled that towering wall of blue-green ice as one of the outstanding sights of the Rockies.

Confidently expecting to take at last the pictures which had eluded him on his first trip, Harmon's disappointment may be imagined when he discovered that the great ice wall, with its deep, ghostly lighted caverns, had disappeared entirely. In its place there ran down to the flats of the Sunwapta a slope of frozen snow so smooth and unbroken that one might have wheeled a baby carriage across it in any direction.

Meltage, with its consequent recession, had played the queer prank, leaving the mighty Athabasca with the flattened snout of a shovel-nosed shark—a thing as devoid of pictorial possibilities as an alkali flat in the middle of the Colorado Desert.

Harmon, who felt that he had been tricked, was so much cast down that it took some diplomacy on "Soapy's" part to dissuade him from packing up forthwith and starting off over Wilcox Pass for the lower Sunwapta. He finally acquiesced reluctantly in remaining over for a couple of days to give the horses a chance to recuperate from the punishment received in crossing the glacier. Before that time limit had expired the witchery of Athabasca Glacier lighting had thrown its spell over the susceptible veteran and he was planning shots faster than he could load film into his camera. One of these was of the scene, previously alluded to, in which old "Soapy's" avid interest in natural history was responsible for the ruin of a shot of incomparable promise.

We had set up with a little gem of a jade green glacial lake in the foreground, the tumbling *serracs* of the Athabasca Glacier beyond, and the glittering snowcaps along the rim of the great Columbia Icefield in the distance. The rays of the low-hanging afternoon sun slanted across the glacier and transformed the lake into a veritable opal of sparkling iridescence. That whole world of ice and snow was bathed in that softly actinic golden glow which precedes by scant minutes the suffusing rose-pink that can be captured by the eye but never transferred to the film.

With only a few minutes left before the glory of the golden flood was quenched in the non-actinic glows preceding sunset, the film jammed twice in the cogs of a camera which had come in for scarcely less punishment than the radio in crossing the Saskatchewan Glacier. At each failure the three packers, who, to give life and action to the shot, were to come down across the surface of the glacier and peer into the depths of the lake from the brink of the ice wall opposite the camera, were sent back to do it all over again.

Tearing out the wads of ruined film, Harmon started the recalcitrant celluloid through the camera for the third time. Finally the temperamental yellow strip crooned its even clickity-clack song again and, literally riding the crest of the last of that glowing wave of light, the men started down over the ice in response to my signal.

The thing was sheer perfection—until the leader of the trio faltered in his stride, stopped, scooped up something from the ice and called his companions over to look at it. Unmoved alike by Harmon's despairing gestures or my Comanche yells, the little group crowded in garrulous wonder around their discovery. When we had yelled ourselves speechless and sunk into silent dejection, the sounding board of the ice threw over to us the drawling accents of old "Soapy," sepulchral and solemn, as if delivering a funeral oration.

"Do youse guys reckon," he queried, "that this yere critter ambled out on the hoof or that he was dropped by a soarin' eagle?"

Further listening-in informed us that our company of movie actors were marvelling as to why the skeleton of a rock-mouse should be found a quarter of a mile out on the ice of the glacier. It was a fortunate dispensation that neither Harmon nor myself was armed.

Of course we made a fourth attempt at a shot, succeeding with it after a fashion. But the glory of that great golden wave of light rolling down across the surface of the glacier, like the perfect melody of "The Lost Chord," was lost beyond recapture.

This same little glacial lake offered opportunity for an interesting experiment with the radio. This was to string the aerial right across a short arm of the milky-green water, with the lead running down to where the receiving set was perched in a niche chopped out of the side of an iceberg stranded on the beach. The ground was buried in the heart of another berg that was lodged against the rocks but with its base still in the water. Baby bergs, well cushioned with folded coats and blankets, made convenient seats for the listeners.

Morse signals, with a great kick behind them, began coming in at once. No one in the party, unfortunately, was able to read them. Though direction (there was only one way the antenna could be strung) favoured reception from the Pacific Coast, many eastern stations could be distinguished as the afternoon lengthened and the pall of darkness swept westward across the continent from the Atlantic.

We had still a couple of hours of daylight left when an Eastern station, which we could not positively identify but which had two or three of the letters WEAF, reported a very hot and oppressive day in New York, with many prostrations and several deaths. There was a great exodus to the seaside in prospect for the weekend.

A little later KGO informed us that an early afternoon temperature of 112 degrees in the shade had been reported at Fresno, with 120 at Yuma, Arizona. Evidently the Pacific Southwest, not to be outdone by the effete East, was staging a little heat carnival of its own.

But what a time and place it was to get reports of heat prostrations and the hegiras of sweltering mobs to the beaches! And how we grinned back and forth about it, to the accompaniment of expressive pantomime, across the magic box! And it really was very amusing and strange and wonderful. Except for an attenuated strip of berg-battered shale under our feet and a pile of glacier-ground rocks just behind us, nothing whatever met our eyes in any direction save ice and snow and a glassy sheet of half-frozen water. Sunstrokes and heat prostrations!—and going on even as we perched there on the ice.

It was unfortunate that gathering clouds, beginning to pour down across the glacier with a threat of storm, made it seem advisable to take down our ice-strung aerial and hurry the radio back to the shelter of the tepee. Quality and clearness of reception was improving greatly as night approached, and it would have been very interesting to carry the experiment on into the more favourable hours of darkness.

In spite of the direction of the antenna, which pointed about south-southwest, Eastern stations were coming in more clearly than we were ever to have them again by daylight. The ice may have had something to do with this, though I am inclined to believe that the extremely careful set-up was the main factor. Most of our subsequent daylight string-ups were made very hurriedly and roughly, many of them on the trail, as when we tried the results of a World Series game and know who won the pool for the day.

We never ceased to speculate, during the four days we remained at our present camp, as to whether or not it would have been possible to head northwest across the heart of the Columbia Icefield and come off by the Athabasca Glacier as we had done by the Saskatchewan. From what we had seen with our glasses from the summit of Castleguard, there was little doubt that the gently sloping surface of the great *mer de glace* itself would have presented few if any difficulties to the pack train. Nor was there, moreover, any question at all that the horses could have traversed the lower two miles of the Athabasca Glacier with much greater safety than they had that part of the Saskatchewan.

That reduced the doubtful zone to two or three miles of tumbling *serracs*—masses of down-pouring ice like frozen waterfalls—between the brink of the main icefield and the comparatively smooth lower glacier. From below we fancied that we could discern practicable passages for the horses down every one of these three or four barriers. From the vantage gained by a two-mile climb up the ice from the base of the flattened snout things did not look so favourable. The ice ridges forming the only feasible route down the first *serrac* were so narrow that it looked as if the horses would have the greatest difficulty in keeping out of the flanking crevasses. And these would have undoubtedly have proved worse at close range. That is the way with ice.

Our conclusion was that, while there was a bare chance that the traverse by this route might be possible, the odds against it were far too great to justify taking the risk. The penalty of failing to find a way over the broken ice at the head of Athabasca Glacier would inevitably mean a night on the main icefield with exhausted and unfed horses. Most of these could probably be taken back to Castleguard Valley the following day—if no accident occurred. I can hardly conceive, however, that a pack train could be disentangled from the mazes of up-tossed ice on upper Athabasca Glacier without serious losses.

The dogs were allowed to accompany us on our climb up the Athabasca Glacier, and shortly showed their appreciation of the unusual privilege by getting lost and cut off from us in a maze of crevasses. No one saw just how the thing happened, but as "Buster" was always the leader of forays far afield, there is little doubt the responsibility was his.

Instead of using his abnormally keen and active intelligence to work a way out of the trouble, however, our prize movie performer, after several futile dashes to all parts of the compass, simply sat down on his haunches, pointed his nose heavenward and howled his despair after the fashion of his wolfish progenitors. La Casse, who could usually control and direct the movements of his pet with the turn of a hand, wig-wagged and shouted in vain. "Buster," with a fighting courage that would send him leaping at the throat of anything from a wolverine to a grizzly, was completely cowed by the impassable depths of yawning green caverns which appeared to hem him in on every side. Possibly, like some humans, he was gifted with too much imagination.

Not so "Tip." Turning his tail to a discredited leader, he began running with that swift, purposeful lope that had carried him around the bend of the Alexandra to an unseen ford, and over the great cataract below Castleguard to avoid a crossing which he sensed as too rough for him to swim. Turned aside again and again but never back, that persistent lope continued until it had brought the altogether unaccountable little Indian mongrel to his favourite perch on "Soapy's" chaparejoed knees.

"Buster," still too much gripped by despair to use even ordinary canine intelligence, had not even the sense to put his keen nose to "Tip's" trail and follow it out. He continued to wail his grief to the hanging glaciers on Mount Athabasca until La Casse, working inward by "Tip's" claw-marks on the ice, gave him a sound drubbing and showed him the way to freedom.

By far our most encouraging radio achievement to date occurred on the last night but one spent in the camp at Athabasca Glacier. The dependable KGO and the Puget Sound stations in both Washington and Canada had become regular routine by this time. Eastern stations we had picked up at frequent intervals when the weather was favourable but they had not always been clear or easy to hold.

The lid of darkness had been clapped down early, following the onslaught of what gave every indication of being an all-night blizzard blowing straight from the Arctic. This, judging from previous experience, boded ill for radio reception. Not only did atmospheric conditions appear to be highly unfavourable, but the increasing violence of the wind threatened to bring down the aerial itself, as had happened one night in Castleguard.

The discouraging prospect was all but responsible for our not limbering up the set at all for the evening. The radio ardour of Ulus, the cook, however, would not permit him to give best to the storm without a fight. When the California and Washington stations came in but weakly and intermittently at the explorative turning of the dials, it seemed that the forecasts of the pessimists were about to prove justified.

"Soapy" and Harmon went to bed and I prepared to follow suit by blowing up the sag in my sleeping bag. Suddenly a cheery cackle rattled through the tepee—unmistakable laughter even after the distortion of leaking out through headphones clapped to the ears of Rob and Ulus. When the latter beckoned me over to take a unit of his phone, I was just in time to hear a hearty, rollicking voice announce that "Station KDKA, East Pittsburgh, Pennsylvany-eeny-iny-ah," was about to broadcast a program "for the benefit of the ladies of Baghdad."

He didn't expect it would reach them, the announcer said, for even KDKA had never been heard quite that far to the east; but, just the same, he was going to broadcast this program to the ladies of Baghdad. He had heard a lot about the ladies of Baghdad, and was going out there to see them sometime. Then they could not only bag dad, but they could bag him as well. His pants were baggy at the knees already, and he understood baggy pants were quite the thing to be bagged in Baghdad.

After a bit more of this light and airy introductory chatter, a program of Oriental music was put on. It must have been furnished by some sort of Arab troupe, for the skirl of the pipes, the peculiar timbre of the high-keyed voices and the barbaric cadences set it apart at once as distinct from the conventional Broadway "Streets of Cairo" imitation.

After the Oriental numbers were over, KDKA continued with several artists who were evidently well known to the listening-in clientele. The jocular announcer called them all by their first names, usually to the accompaniment of a bit of badinage, as when he stated that "this old broken-down, Ed Squires, will now sing a new song entitled 'My Radio Girl.'" Squires came back with delicate repartee in similar vein, as did most of his fellow artists.

The jolly KDKA "Master of Ceremonies" also poked much innocent fun at the announcers of various rival stations, apparently on the assumption that they were known to all of his listeners-in. One of these shots, the gist of which we did not get, was directed at "my old friend and well-wisher Bill Hays of Hastings." Our acquaintance with the genial Hays had been a somewhat spotted one up to this time, but we subsequently came to know him as a nightly visitor when KFKX changed to a wavelength better suited for penetrating to our mountain retreat.

We held KDKA without a break right on to its signing-off, but missed the exact time of the latter because the dogs were still howling their appreciation—or rather their disapproval—of the concluding soprano solo. From the fact that it was after nine o'clock, Mountain Time, it must have been later than midnight in Pittsburgh. This led us to conclude that the program, which consisted almost entirely of light musical numbers and jazz, was a special after-the-theatre concert or something of the kind. We cut in on it several times later, but never on successive nights.

The reaction of the dogs to the radio was of never-failing interest. The voices leaking out of the headphones were very real to them. Their general behaviour on hearing most of the sounds was similar to that manifested in listening to the sounds of an approaching pack train. They were put on the alert instantly but were not seriously concerned.

The distinct and unmistakable voices of women, however, were quite another matter. Our zealous canine guardians seemed to feel that women had no business about the place, and so actively resented sounds which appeared to herald their approach. In the case of KDKA's soprano their howls of protest were more in sorrow than in anger, but when, half an hour later, CKCD of Vancouver came on with a so-called "Poetess of Passion" reading her own verse, they were just plain fighting mad.

The name of the ethereal invader, as announced, sounded very like "Carry-me-homah," but was probably just Carrie something or other. The mental picture she conjured was of a tall, black-draped figure, swaying before the microphone and boring it with eyes the colour of purple grapes. Her voice was full and resonant; the burning words were winged with the kick of a sharply flexed diaphragm behind every syllable.

It was the militant "Buster" who tore loose the flap of the tepee door in leading the charge out into the snowstorm, with the apparent intention of attacking the invader in the timber; but it was the subtle-minded "Tip" who was the first to sense the futility of a campaign in that direction, and who came yelping back to tear the enemy out of her protecting black box. Here, presently, he was joined by the baffled "Buster," and no spine-backed porcupine cornered in the timber was ever told more plain, unpleasant truths about himself than were voiced by our two faithful guardians in trying to bring home to that elusive poetess of passion just the manner in which she would be chewed to shoe-strings if she ever came out of her hiding place to disturb the Nirvanic calm of our hitherto peaceful camp.

Both dogs took a lot of petting and soothing before they would cease worrying the radio box and settle down in their places by the fire. But so, also, would most people who had to listen to a poetess of passion reading her own verse.

Our last day at Athabasca Glacier we took advantage of bright sunshine and brilliant lighting to climb 1,500 feet to a shoulder of Mount Wilcox to make a panorama of the rim of Columbia Icefield from the northeast. Just as we reached the first of the meadows leading on to Wilcox Pass the dogs started up and gave chase to the finest specimen of black fox I have ever seen. Fully as long as "Buster" though standing not quite as high, his silky hair glistened in the sunlight like a block of newly mined anthracite. A streaming, bushy tail, almost as long as his body, was tipped with a flowing tassel of snowy white.

With a yell of, "There goes 2,000 plunks worth of fur," "Soapy" spurred on his speedy thoroughbred and dashed away in the wake of the flying dogs.

For a mad minute or two it looked like a real race. "Buster" and "Tip," yelping with ecstasy, fairly tore up the grassroots in their eagerness; the swift-footed "Alice," with "Soapy" leaning flat along her extended neck, ran in a pyrotechnic spatter of steel-struck sparks.

Then that animated fur boa, apparently having only tolerated the pursuit as long as it interested him, simply dematerialized. No other word quite describes that lightning disappearance. One moment there was a lazily loping black fox, with two dogs and a horse closing hard upon his leisurely heels; the next two foolish dogs and a swearing man on a reined-in horse were watching a dark streak parting a path through the lush grass and flowers of the mountain meadow.

As usual, both dogs and man, Chinese-fashion, tried to "save face." "Buster" and "Tip," spying a flock of sheep a few hundred yards off on the mountainside, dashed off in that direction in a studied attempt to convey the impression that these comparatively slow-footed denizens of the cliffs were really all that they wanted to catch anyhow. And when the bighorn, finally taking fright, left their pursuers in a few swift leaps, the wily "Buster" promptly turned aside and led a fierce but futile attack upon his perennial enemy, the rock-rabbit.

Old "Soapy" tried to hide his disappointment under a cloak of sunny philosophy, saying the fur would not be prime until midwinter, and that then, maybe, he would snowshoe in and trap both the owner and his mate.

All through the rest of the trip the plan of coming back in January to collect the three or four thousand dollars worth of fur, hitherto denied to Fifth Avenue or Rue de la Paix through the misfortune of remaining on the backs of "that black devil" and his family, was a subject of never-failing interest among the packers. Whether they went or not I have never heard.

Two or three black-fox hides, with those from the incidental trapping of marten and wolverine, would seem to have made the venture well worthwhile—providing, of course, the foxes were caught. This much to be desired consummation, however, the men always seemed to take as practically assured. None of them ever vouchsafed other than negative or hostile notice to me on the several occasions when I interrupted discussions of their plans for spending the money by quoting the beginning of the old cookbook recipe, "First catch your hare...."

The view from the point at which we made our panorama on the shoulder of Mount Wilcox was both beautiful and interesting. The three great ice falls by which the Athabasca Glacier descends from the Columbia Icefield through a gap between Mount Athabasca and the Snow Dome were clearly defined. Here, again, it looked as though a way for the horses could be found down through the mazes of crevasses, but we were really too far distant to make observation on that point really dependable.

My considered advice to any mountain party which might be tempted to chance this extremely fascinating but highly hazardous traverse is just plain "Don't!"

The Dome Glacier, tumbling down between the abrupt northern wall of the Snow Dome and Mount Kitchener, was visible from our vantage for nearly all of its short length. Fallen rocks give it a very dirty, unwashed sort of a look, especially where its snub snout is pushed out against that of the clean white of the Athabasca Glacier at the head of the Sunwapta.

The sheer wall of the Snow Dome is the one I have mentioned in the previous chapter as presenting so sharp a contrast to the smooth, unbroken slope running up from the Columbia Icefield to the rounded summit of that strange mountain.

Baptie, who had spent most of the day rounding up strayed horses, found time in the afternoon to bring down and into camp a fine mountain ram. It was shot on the "Big Hill," where the Banff-ward trail winds down to the North Fork of the Saskatchewan.

Why one bunch of the horses, with so much good grazing near camp, should wander ten miles to the south, was no more to be accounted for than the fact that another bunch climbed the steep slope to Wilcox Pass and were found just as far to the north.

With clear, cold weather locally on our final night at the Athabasca, radio came in better from Pacific Coast stations than from those of the East. This was in direct contrast to reception conditions of the previous night, though no change had been made in the direction of the aerial. Distant storms, doubtless, had much to do with the difference.

KGO came in clear as a bell all evening, including the late supper concert of the Fairmount orchestra in San Francisco. Finding Ulus and Rob singing and swaying in unison to the seductive jazz winged to their ears across a thousand miles or more of mountain and plain, Harmon and I seized the occasion to set off a calcium flare in the tepee campfire and transfer the amusing scene to celluloid. This gave opportunity for the subsequent cut-in of a highly contrasting scene showing the after-theatre dancers jazzing at the Fairmount.

Thus it was, watching our chances as they came—and especially on the score of these little human touches—that we continued to build up our scenic of the Rockies.

Chapter ix
DOWN THE SUNWAPTA

The final base of photographic operations we had planned in completing our work on and around the Columbia Icefield was the head of the Athabasca, immediately under the great peaks of Alberta, The Twins, King Edward and Columbia. As the crow flies, the distance from our camp at Athabasca Glacier was not great, but by the only possible route with pack train it was close to 80 miles of flooded valleys.

Because the canyon of the Sunwapta (which begins three or four miles below the terminal moraine of Athabasca and falls a thousand feet or more between closely boxed walls) was impassable for horses, we had to begin the long and circuitous journey by climbing over the lofty Wilcox Pass to the east.

Wilcox Pass was named from W.D. Wilcox, who, with R.L. Barrett, was the first white man to use it when he crossed, in 1896, from the North Fork of the Saskatchewan to the Sunwapta on his way to Fortress Lake. There are one or two higher passes in the Canadian Rockies, and many that are rougher, rockier and more difficult to traverse in good weather. It is its most unbrokenly bad weather that has gained for Wilcox so sinister a reputation. It is the first to be closed by winter snows and the last to be opened by spring thaws. This is due less to its altitude than to the fact that, with the Columbia Icefield acting as a great condenser of the moisture-laden airs blowing in from the Pacific, the region has what is probably the heaviest snow and rain falls in all the Rockies.

Of the dozen or so parties that have used the Wilcox Pass route in the last decade, most of those going over very much earlier or later than midsummer have had snow trouble.

Our party, late as the season was, had an exceptionally fine day for the lofty traverse. There was a hard, tiring climb up from the south, several miles through rolling mountain meadows where our worst difficulty was to keep the horses from scattering to graze, and then a steep descent to through knee-deep moss to a comfortable camp a mile below timberline.

While Harmon and I clambered back to the heights with the cameras in the hope of getting a chance for mountain sheep pictures, Ulus and Rob, with several hours of daylight to work in, set about making what they announced would be the most perfect string-up of the antenna we had yet had. The aerial was the important thing, they explained; once get that right, and only the sky and the oceans were our limit.

It was twilight when we returned to camp, and quite dark before supper was over. Our experts, gleefully telling how the antenna was stretched between a cliff and the highest trees we had yet found, opened up the radio and began tuning in.

It was a gala night program promised for that evening by KFI of Los Angeles that they were especially anxious to get, but if a few new Eastern stations wanted to come through and entertain us, so much the better. We would give them all a turn—provided, of course, they had the power to reach us. It was juice that talked in this long distance stuff. Thus Ulus.

After getting up our hopes like that, it was, naturally, a bit disappointing to find we were cocking receptive ears into empty air. The strongest sounds heard were the magnified clangs set ringing when an impatiently jiggered needle reached the end of its beat.

Not without experience of the reactions of French-Canadian temperament, and especially of a French-Canadian cook, Harmon, as Ulus' face grew redder and redder with suppressed rage at the misbehaviour of his pet, rose quietly and left the tepee. Trailing his bobbing flashlight a few minutes later, I found him feverishly engaged in hiding the axes and other instruments of destruction for fear the irate cook would use them on the radio.

The dark, sweeping line of the antenna was outlined clearly against the cloudless sky as we turned to go back to the tepee. Harmon, his eyes fixed on the polar star, stopped short in his tracks and then broke into a chuckle.

"That is sure a wonderful string-up the boys made," he whispered. "Only thing wrong seems to be that they pointed it off toward Labrador."

And that, beyond a doubt, was what was at the bottom of the unexpected silence of the ether. The lead has been brought down from the northern end of the antenna. We were careful not to repeat the mistake, and that was the last time the air refused to talk to us.

Off to an early start the following morning, a mile through timber where the way had frequently to be cleared with an axe took us to the brink of an extremely steep descent of over a thousand feet to the Sunwapta. The first 300 feet of this was down a grassy, untimbered slope where even the mountain sheep had found it easier to keep their footing by making long zigzags.

If there had been only one of these sheep trails it might have been possible to keep the horses to it. With half a dozen of them to choose between they scattered, and so did the packs of two venturesome animals who plumped for perpendicular rather than zigzag descents. One of these was "Wolverine," with the radio, who described two perfect parabolic-curved somersaults, and half a dozen that were not so perfect.

At the foot of the open slope the flying mass of horse and pack telescoped against a sturdy spruce trunk as neatly and compactly as the lily in Tennyson's song that

> ... folds all its sweetness up
> And sinks into the bosom of the lake.

We had long since become deeply entrenched in the belief that both the "Wolverine" and the radio carried their own protecting magic. The squat, dachshund-like pack horse, when we finally got him on his feet and smoothed out the bends and creases, was found to be little altered in shape and not at all in the ability to function. Neither was the radio, as we discovered when we set it up two nights later.

The radio was riding much better on the trail after the overhauling given it at the camp at Athabasca Glacier. Cutting down the strips of cedar salvaged from the crushing on Saskatchewan Glacier, we had constructed a new box that was four inches less in width than the original. It took a deal of fitting before the block of batteries was humoured into a shape that would go into the reconditioned container, but once there they rode much more solidly and compactly than before.

Better still was the fact that the narrower box missed a good many trees and ledges that the broader one had tried to push out of the way. It also came more nearly to a balance with the grub box which rode on the other side, making a load much less likely to turn or press unevenly.

With little in the way of a track to follow, the pack train spread out a good deal in working down the last thousand feet of descent through the timber. As there was not much of any place to get to but the river whichever way they went, the only effect of the "line abreast" formation was to make a lot of extra work in extricating horses from the maze of deadfalls in which they were constantly becoming entangled.

When a pack train is going ahead in single file the cutting away of an obstructing tree will allow all the animals to pass. If they scatter in bunches through the timber a way has to be cleared for each line. This breaking up of a train is, of course, avoided as far as possible. It is always liable to occur, however, in working down a steep slope through heavy timber, especially where there is no previous track to follow.

We came back to the Sunwapta at the foot of the deep canyon leading down from the flats below Athabasca Glacier. It was flowing in a single boulder-choked channel that was beset with many swift chutes and tumbling rapids.

As the mountain walls fell back and the valley widened the descent of the stream was less tumultuous. At the end of a mile it was spreading over gravel flats and occasionally dividing into broad, shallow channels which gave no trouble in fording.

The pack train rattled along at a fast pace down the broadening flats, crossing the river at will to shorten distance. Everything appeared favourable to keeping right on to the mouth of the Pobokton, where the maps indicated the first stretch of good trail to be encountered since leaving Lake Louise.

The rim of the northerly extension of the Columbia Icefield, bright with sun-glazed snow-crowns and sparkling glaciers, formed the skyline to the west. Milk-white glacial streams came pouring down at frequent intervals to augment rapidly the volume of the main river. The latter, fortunately, spreading rather than deepening as its flow increased, was still easy to cross for a number of miles.

What at a distance had appeared to be a yellow-brown boulder lying on a gravel bar, provided us a lively diversion when it suddenly doubled in height and began to move. With a shout of "Grizzly!" "Soapy" began calling off the dogs (neither of which were "bear broke") and fumbling for his rifle.

The dogs were quite out of control, however, and with the belligerent "Buster" yelping in the lead, dashed in for a death-grapple. Expecting every instant to see a paw-cuffed canine rise in a widening parabola against the glacial background beyond, all five of us left the pack train to shift for itself and spurred on to the rescue of our beloved pets. If they couldn't be saved, we could at least wreak revenge on their murderer and give the victims a Christian burial.

Fond thoughts of "Buster's" lordly but engaging arrogance, of "Tip's" bright and winning ways, arose in my mind as I gave swift "La Belle" her head and tunnelled into the wake of gravel spurned by the hoofs of the speedy "Alice R." Every man-jack of us subsequently confessed that he was thinking of the dogs as among the dear dead departed right up to the finish of that whirlwind race across the flats.

With all three combatants rolling in an inextricably entangled ball of milling legs and flying fur, it appeared that we had come too late for a rescue and that revenge and burial would have to be the order of the day. But as we reined in our horses on the verge of the fray it became evident that there was still a bit of the vital spark left in the two viciously snarly and biting knots of kinetic energy which, a minute before, had been our romping, affectionate pets.

Yes, there was a lot of life in our good dogs yet; really very much more than in the mauled and bedraggled carcass of the ancient billy goat they were worrying to a death that would have been but a matter of a few days in any event.

With teeth worn to the gums or gone entirely, and with tottering legs no longer strong enough to carry him over the rocks after grass, the unfortunate old patriarch of a flock, ceasing to challenge the law of gravity among the cliffs, had descended to the warm river flats to die alone. Dragging his emaciated frame about the clay-stained rocks had muddied the once snowy hair to the yellow-brown which, seen at a distance, had led even the eagle-eyed Rob to second "Soapy's" snap verdict of "Grizzly!"

With the end already near, it was really very much of a mercy to have had a slow death by starvation replaced by the sudden surcease winged by a mushrooming soft-nosed bullet. Doubtless the brave old patriarch would have preferred to have it that way himself, though it is a pity his pride could not have been spared the humiliation and ignominy of that mauling by the dogs.

For four or five miles the river broadened and shallowed in widening flats, the current growing slower and slower as the declivity decreased. A light stratum of soil above the gravel brought grass with it, and presently blooming flowers.

As the layer of silt increased in thickness the river courses multiplied and burrowed, until we found ourselves working down across a network of channels, many of which were too deep and steep of bank to cross without danger of wetting the packs. That made it necessary to take circuitous courses around the bends and over the jutting points of mountain reaching out like capes into the sea of the level flats. Our swift, direct progress down the upper valley gave place to the slow, wearing windings which had proved so troublesome and costly on the North Fork of the Saskatchewan and the lower Alexandra.

For a while—with the silt fairly solid underfoot—we were more annoyed than concerned over the delays imposed by the changed conditions. Then, swift and sudden as a thunderclap, came trouble that was near to disaster.

La Casse was leading the pack train around the sharp cutbank at the outside of the bend of a channel too deep to risk fording. Baptie was midway of the straggling string, with "Soapy" bringing up the rear. Harmon and I, stopping now and then to take pictures, had dropped several hundred yards behind "Soapy."

A wild yell—more in anger than in anguish—from La Casse, the staccato snortings of frightened horses and an obbligato of excited canine yelps, constituted the only prelude to the sudden transformation which followed. One moment there was a pack train of normal four-legged horses winding quietly around the bend; the next ten of them had been reduced to floundering, legless trunks.

The five leaders, including the mount of La Casse, had started the show by going belly deep into the mud as suddenly as though let through trap doors. This was too much for the wily "Rat," who promptly went over the bank into the river, followed by the amphibious "Nelly," "Wolverine," with the radio, and the three next in line.

Baptie and La Casse, with their own horses down, were helpless for the moment; Harmon and I were out of the picture entirely. So there was no one to divide the credit with old "Soapy" for the swift dash which resulted in halting midway the dive of "Jerry," the movie horse, for the river, and the subsequent turning back of the three or four packs which followed.

That saved the movie camera but not much else. As Baptie dismounted to give his horse a chance to regain its footing, the excited animal jerked away from him and floundered over the bank after "The Rat" and his aquatic followers. La Casse managed to hold his horse on the verge, but the four pack animals immediately behind him rolled over into the river as fast as they freed themselves from the mud.

There must have been nine or ten feet of lazily flowing water under the caving bank, so that every horse, at the end of the five-foot drop from above, had enough momentum to carry him, pack and head, well under the surface. When they came up they had lost all sense of direction, and so kept milling round in bunches and trying to climb on each other's backs. This was the one thing needed to make the soaking through and through of the packs quite complete. It was especially annoying in the light of the fact that there was an easily sloping and fairly solid sandbar over which to get out of the river on the opposite side.

The inevitable toll of all kinds of provisions save canned goods was serious enough, but the threat of real disaster was in the fact that all of the photographic film—exposed and unexposed—was in the boxes on the back of an especially enthusiastic swimmer called "Nig." Getting him out and unpacked was the first thing to be done. This could only be accomplished from the farther side.

La Casse, with Baptie clinging to the tail of his swimming horse, crossed below the bend. The rest of us found a shallower ford above, where the movie horse could wade without danger of wetting his pack. As soon as the swimmers saw us come out onto the solid ground beyond the east bank, all but two of them wallowed over to join us of their own accord.

It seemed just a shade beyond routine hard luck when it transpired that the two absentees were "Nig," with the films, and "Nelly," with the scarcely less precious salt and sugar remnants. The fault appeared to have been that of the playful trail-born filly. Trying to hurdle "Nig's" back, the water-slacked lash-rope had caught over the end of a pack box, and there she hung, like a rearing bronze horse on a fountain, enveloped in the showers of spray kicked up by her flying hoofs.

The lively water spectacle would really have been very funny if only the two actors had been laden with packs of less vital importance, such as Harmon's tent or "Soapy's" bedroll. Now it was "Nelly" that was beneath the surface, now "Nig"; and both of them seemed to assume it was the other that was responsible for the ducking. It was the high-strung colt that was the maddest about it, and some of her bites into "Nig's" glossy black mane were a bit more than playful love nips.

As there was no way to reach the floundering equine Noyades to disentangle them where they were, Baptie probably hit upon the best solution of the problem when he lassoed the both in one loop and so made it possible for us to tail on and bring them over to the bar with a "yo-heave-ho" on the riata.

Though there was no favourable camping ground for many miles, the necessity of salvaging everything possible from the soaked packs made a halt where we were imperative. Harmon and I began on the pack boxes containing the photographic supplies the instant the "Nelly"–"Nig" liaison had been dissolved. The men took the rest of the outfit a quarter-mile across the flats to unpack and make such camp as was possible on the fire-scarred eastern mountainside.

There was mud and water in both pack boxes, with the cardboard containers of all the films reduced to soft masses of pulp. Moving-picture film, together with the roll film for my still cameras, appeared uninjured in their tins. In fact, the only loss suffered by the tinned film was a single roll of my own, to which water had penetrated under the loosely taped top of its can. Not a foot of the movie film was hurt by water, either at this time or later.

The case of Harmon's large stock of film packs, however, appeared more serious. With the fibre outer wrappings turned to slush, there still remained coverings of oiled paper and tinfoil. Most of the transparent, ruby-coloured paper had drops of water standing upon its oily surface, and a bit of this had gone through to wet the layer of foil. But the black paper of the pack itself was hardly more than moist on the outside, indicating that the sensitized film beneath was almost certainly unaffected by the bath. This was borne out by subsequent development in Banff, many weeks later.

It was necessary, of course, to get rid of all the moisture both in the film packs and their wrappings before packing them up again, and on that tedious job Harmon and I spent all of the rest of the day. A fly was erected to break the direct rays of the sun, and under this the unwrapped film packs were laid out on tarpaulins to dry in the wind. The last of the little rectangles of film was not rewrapped until after dark.

It was a great relief to all of the party to have assurance that no injury had been done to the unprotected film packs. Harmon really cared more for these still negatives than for the movie films, for they were to complete a collection of Canadian Rockies views which had been 20 years in the making.

The loss to provisions was a heavy one. The salvage from the sugar and salt amounted to only enough of each to fill a couple of empty baking powder cans. Worse still, some of the dirty brown chips from each lot had become mixed in drying, so that one had to taste them to be sure of what he was getting. And if one forgot that explorative tongue-dab, the consequence was liable to be a salted cup of coffee or a sugared chop of mountain mutton.

Flour and oatmeal suffered heavily, especially the latter. Flour automatically forms an indurated protective coat at the touch of water, leaving as a net loss only the layer that hardens. Oatmeal will do the same thing if packed tightly in its bag, but the exclusion of water from the inner heart is less complete. Our oats, being loosely sacked, took water all the way through and so left little salvage.

The remnants of our dwindling stock of canned stuff were, of course, not hurt. Neither was the dried fruit nor the dehydrated vegetables, though not a whole afternoon of sunning would reduce either to its former concentrated bulk. This was of little importance now, however, as there was pack room and to spare.

The radio we drained, cleaned and dried as best we could, but the wretched camp offered no trees high enough to carry the aerial even if there had been anyone with enough time to string it.

All of the beds were wet; that of Ulus was soaked. The unlucky cook poured a gallon or more of water out of his sleeping bag but never got it thoroughly dried until toward the end of our long vigil under Mount Columbia.

That was another "maiden" camp to name, and we named it—thoroughly. Suggestions came in all afternoon and most of the night. The most refined of them was much worse than the worst suggested for our flood camp on the lower Alexandra. And even those, as I have already told, were unprintable save on asbestos.

With wet beds and the floor of the tepee tilted to a 20-degree slope, no one doubted that we were going to have a miserable night of it. We did. An individual anthill under three of the four beds, with "Buster's" sleeping blanket thrown over a fourth nest of the vicious little pests, lent a final and fitting "sorrow's crown of sorrow" to the end of a far from perfect day.

Anxious to put the accursed place behind us as speedily as possible, we rolled out before sunrise in the hope of rounding up the horses early and making a start that would compensate for the time lost the previous day. With the mountainside blocked with deadfalls, none of the horses had wandered far during the night, and we had all but one of them in and under saddle before breakfast was ready.

But just as the strength of a chain is that of its weakest link, so is the ability of a pack train to get under way dependent upon the last of its units to be found and brought to camp. This lone stray of ours—a nameless sorrel—was still adrift at eleven o'clock. It was Harmon who finally tracked it down, just before noon, where it was held so firmly in a natural stanchion of deadfalls that the close-gripped neck could not move far enough to tinkle its bell.

A midday start, of course, meant another short stage, even without further recurrence of trail trouble.

The valley floor began to tilt downward more rapidly a mile below camp, and from there on the improved drainage of the flats made for less mud. There were difficulties of one kind or another right along, but no serious obstacle appeared until we were brought up short at the upper side of a great slide of broken rock thrown all the way across the valley—probably the prank of an ancient glacier.

With the unfordable lake dammed back where the slide blocked the river barring a detour to the right, and the mountainside appearing too rough and steep to allow passage to the left, there seemed nothing to do but try to take the pack train across that 100-yard-wide barrier of sharp-edged, up-tossed boulders. "Soapy" shook his head dubiously, and then began building a trail by carrying small rocks and throwing them in the cracks between the big ones. A solid blocking of the worst of these holes was the only possible way in which the horses could be prevented from slipping through and breaking their legs.

Harmon, setting up his tripod, prepared to reap what compensation he could from adversity by making a movie shot of the precarious passage. There was no question about its making wonderful action Stuff, but inevitably at a heavy price. But since the price would have to be paid anyway, it was only good business to make a record of the transaction.

After an hour's work carrying and fitting rocks had made it evident that the building of even the most primitive trail across the slide might well take us the rest of the day, "Soapy" gave up and announced that it would be easier to cut a way through the timber of the western mountainside. We were confirmed in this decision when dips into the close-growing timber at several points revealed unmistakable signs of the previous passage of a pack train. Cut logs, fragments of pack covers and horsehair rubbed into the bark of trees told the story of a hard but apparently successful fight. Many trees which had fallen since the former traverse would have to be cut out, but this was nothing compared to the labour and the risk of venturing over the rock slide.

We bogged several horses in crossing the overflowed flat forming the western section of the floor of the valley, but after that it was just the slow, tedious labour of clearing a way ahead. This was no worse than the work we had had on the slope above the flooded lower

Alexandra, and the horses—with lighter packs—were now under better control. The whole outfit was down into the open below the slide in less than an hour.

The river, which, after being dammed back by the slide above, came tumbling down over that obstruction in a noisy cascade, now flowed in a narrow, boulder-choked channel too dangerous to ford. This was unfortunate, as the eastern bank appeared to offer much easier and more open going than the narrow bench to which our line of progress was restricted. Our map, indeed, even indicated that there was a trail leading down the Sunwapta from Jonas Creek.

When our glasses revealed blazes on occasional trees 50 yards back from the farther bank it appeared probable that there might actually be some sort of path there.

When the mountain wall advanced to meet the river a mile below the slide, no alternative remained but that of fording. Picking the least unfavourable place with much care, we put the pack train in where a muffled "chunkity-chunk" from the swirling depths of the milky glacier water told of boulders complaining because they could not find a resting place in the impetuous current.

Everything considered, I am inclined to rate this as our most dangerous crossing, not excluding the two fords among the rolling boulders of Castleguard River. The water, lapping but a few inches above the bellies of the horses on the lower side, almost surged over the backs where it struck them from above. This terrific pressure, and the fact that the staggering animals had to prod with their hoofs at every step between boulders which were rolling or in precarious balance, conspired to make the passage a matter of touch-and-go from bank to bank.

As had previously happened on Saskatchewan Glacier, the sense of really grave danger seemed to make the horses forget their cussedness and give the best that was in them to the serious work in hand. Carelessness or stubbornness had been responsible for their being swept down in many an easier ford than this one. Now, however, something seemed to make them comprehend that the loss of footing in that wild welter of foam and rolled boulders would be fatal—that nothing could save them from bumping down half a mile of rapids that were near cascades.

And so we crossed without a single horse losing the footing that would have been the prelude to disaster. I felt the spirited "Belle" quivering under me as she fumbled with an explorative hoof for every step. But not once did she balk, or even hesitate long enough to throw confusion into the tense string behind.

It was with real relief that we watched the last horse clamber up the bank and shake the water out of dripping tail and mane. One bad stumble would have meant that the recovery of even the pack was up to our finding the body of the animal which had borne it. Nor is a roll among boulders, tangled up among the legs of his floundering mount, a thing lightly to be courted even by a man. One should be careful, of course, to ride with his feet free during a bad crossing; but a roll among the rocks of a rapid with a kicking horse is not a pleasure jaunt, even without the complication of being linked by a stirrup-jammed foot and left to fight it out like two cats over a clothesline. Kicking is one of the several things that a horse does better than a man.

We were a good deal concerned about the dogs at this crossing. There was no chance of their being able to stick on a swaying pack in such rough going; nor was any of the five of us likely to have a free arm to crook around a canine passenger, as I had done at the Saskatchewan with "Tip." There was nothing left but for them to fight it out, "every dog for himself and the river take the hindmost," slightly to paraphrase the ancient saying.

The game pair surely made a brave swim of it, especially "Tip," who had been endowed with nothing comparable to "Buster's" formidable equipment of strength and courage. Jumping in far up on the western bank, they were swept down through the line of the wallowing horses—fortunately, without being trampled underfoot—finally to bring up in an eddy lapping the eastern bank a hundred yards below.

That narrow run of back-current came in a very convenient place. Carried beyond the rocky ledge which formed it, our pets must have been drawn into a series of chutes and whirlpools where their fate would have been considerably in doubt. "Buster," I am inclined to believe, would finally have crawled out with his flaming life-spark still unquenched, but more likely than not on the opposite side of an unbroken run of rapids, where there would have been no way of reaching him.

"Tip," less robust of heart and body than the husky, would have needed another "brain-wave" to bring him out alive.

With the day already far advanced, we decided to camp at the first favourable site that offered, leaving a search for the trail to go over to the morrow. A half-hour later we were setting up the tepee on the grassy, timbered island, formed by the splitting of the river into two rocky channels by a huge logjam.

The dogs charged out through the tepee door several times during the night, but, with the roar of the rapid rising from either side, we gave little heed to their violent spasms of barking. In the morning we found the tracks of four deer, two caribou and a moose pressed into the plastic mud of the lower lip of the island. The caribou had plunged into the main channel of the river—probably frightened by the onslaught of the dogs—and swum to the opposite bank. With a ten-mile current rushing down the flooded stream, it must have taken a vigorous bit of pawing to land above the under-suck of a jutting logjam 50 yards below. I should never have dared to work on so narrow a margin in swimming a horse.

Packed up for an early start in the morning, a 50-yard clamber up the steep side of the bench flanking the eastern side of the valley brought us to a broad, even swath cut out through the timber. Boggy in many spots and plainly very little travelled, this was still the first thing worthy of the name of trail we had touched since leaving Lake Louise. The trees had been cut and thrown back out of the way, there were "corduroys" of well-fitted logs at the worst seepages, and, running along on one side, was the wire of what had once been a telephone line.

Pushing the horses at a brisk pace along what seemed (by comparison with what we had been traversing) almost a paved highway, we reached the Sunwapta ranger's station early in the afternoon. The cabin appeared to have been unoccupied for some time, but an entry in the register disclosed the fact that it had been visited, a fortnight previously, by a party which had been climbing mountains near the head of the Athabasca. Howard Palmer, engineer, author and alpinist, and Conrad Kain, Austrian guide, were among the well-known names on the pencilled list.

Definite word that an outfit had been in to the head of the Athabasca so recently was good news for us. It meant that the route had been traversed at least once in both directions, and that much cutting and clearing, which otherwise would have fallen to our own axes, had already been done.

We were sorry to have missed foregathering with so notable a party of mountaineers. Howard Palmer, who has climbed much in and mapped considerable areas of the Selkirks, I had known through correspondence since my voyage down the Big Bend of the Columbia in 1920. Conrad Kain, a mountain companion of Harmon's for many seasons, I had met at the Lake of the Hanging Glacier just before starting down the Columbia. It was he, indeed, who pitched over the cliff the frozen goat which came so near to wiping Harmon off the face of Horse Thief Glacier, as related in the chapter on scenic work.

Two hundred yards below the rangers' station, where a sign marked "To Fortress Lake" pointed toward a straggling path disappearing up a rocky slope, we turned our backs on the open white man's trail which ran on to Jasper, and resumed once more the way that had been blazed by the Indian.

Crossing the gorge of the Sunwapta on a crudely but solidly built log bridge picturesquely located just below a 30-foot fall, we climbed a low divide and descended, through the close-growing timber clothing the slopes of a series of terrace-like benches, to the banks of the Athabasca. Two miles farther upstream, where a narrow neck of alluvial deposit separated a broad overflow lake from the flood-high river, we halted for the night. Marks on rotting fragments of pack boxes indicated that a government surveying party had camped on the same spot many years before.

Chapter X
UP AND DOWN THE ATHABASCA

ON THE ATHABASCA WE WERE back again close under the eaves of the roof of the Continental Divide. Waters from the glaciers crowning the lofty heights above our camp drained both to the Pacific and to the Arctic. Towering against the northwestern skyline were the snows of the historic peaks of Brown and Hooker, twin sentinels of old Athabasca Pass. Still farther north, but out of our present line of vision, was the fine summit recently named Edith Cavell. Black Monks and Chaba Peak were to the southwest of us, with Columbia, Alberta, King Edward, The Twins, and others of the great group of peaks flanking the Columbia Icefield on the north, waiting to appear as we pushed south up the Athabasca.

As our initial camp on the Athabasca was located at the first point yet touched to which there was a practical certainty of our subsequent return, we took advantage of the opportunity to leave there everything in the outfit not likely to be needed on our journey to the head of the river. This consisted mostly of exposed film and heavy clothing brought along for the return journey to Banff through the early winter snows. Made into a compact bundle and suspended by spare radio antenna wire from the apex of a stout tripod, "Soapy" declared the cache was proof "'gainst everything but wolverines, and mebbe him."

If there were provisions in the bundle, and especially bacon or other meat, the packers said it would only be a question of time until a savage and resourceful wolverine would contrive to worry off the wire and rip the contents wide open. As food was the one thing above all others that we could not afford to leave behind, there was no chance that predatory prowlers would be attracted by any odour more tempting than that of celluloid.

With the river, swollen by the meltage incident to another succession of unseasonably hot days, out of its banks and spreading over the flats, we had wet though not especially boggy going all the way to the mouth of the Chaba. This was about four miles.

Above the Chaba, where the route to Fortress Lake branches off toward the west, the valley of the Athabasca narrows almost to a canyon, with the river roaring down it in an unbroken succession of rough rapids. Along here we kept to the east bank, following as well as we could the trail traversed and partially cleared by the mountain-climbing party a few weeks earlier. This led most of the way over benches black with burned deadfalls. Descent to the flooded river level was only made where a cliff or impassable ravine blocked the way over the more solid ground above.

Most of our trouble came in fording the narrow, deep mouths of torrents pouring down from the mountainside. Boulder-choked and steep-banked, with the murky glacial water making it impossible to judge depth or the character of bottom, these harmless-looking holes proved dangerous pitfalls for horses. Packs had repeatedly to be removed in extricating mired animals, and both broken backs and broken legs were always imminent possibilities.

With both human and equine members of the outfit thoroughly exhausted from the rough, punishing work, we were glad to make early camp at the first point reached where there was grass for the horses and a level spot for the tepee.

We were on the way early the next morning, hopeful of reaching the site of our base-camp before nightfall. A mile of pulling and hauling through the usual rocks and fallen timber brought us out to a point where the valley widened to half a mile or more, and here we welcomed the chance to head straight up the flats toward our destination.

The swollen river, sweeping across the floor of the valley in wide curves, blocked our way repeatedly. The current was deep and swift, but—with a solid bottom of gravel and small rocks underfoot—never dangerous. Only once were the horses swimming at a ford, and that was for a very short distance. Progress was rapid and steady from the moment we broke free from the clutches of the entangling timber and narrowed our problem down to fighting it out with the river in the open.

Just as when nearing the Columbia *mer de glace* by the Alexandra, we again found ourselves entering a kingdom of ice and snow. These were only in evidence afar, however. Where we toiled up the hot valley floor water was the dominant element. Waterfalls, sparkling like diamond necklaces, ringed every jutting headland; cascades, churned white as milk, streaked every slope. We were at the climax of the last of the several great thaws of a remarkable summer.

Mount Columbia, the tip of the snowy pinnacle of which we had seen for a moment in crossing a high bench above the Chaba the previous afternoon, played hide-and-seek with us all day. This was traditional Columbia tactics, Harmon explained—it had been pulling this "now-you-see-me-and-now-you-don't" stuff ever since it was discovered. The trick did not worry him much, he said, as he was quite prepared to employ his familiar siege strategy—that of sitting down and waiting for a complete capitulation.

We were riding side by side up the valley when Harmon made this confident announcement of policy, so there was really no wood to touch to propitiate the ever-jealous Master of Futurities. Possibly Harmon had never been sufficiently chastened to learn that one ought never to omit doing that sort of thing. I muttered a pious Mohammedan "Insh-allah," but even that could avail naught for my rash companion. The time was near when he had to learn by bitter experience that it is not well to fling challenges for a bout at waiting in the teeth of such a practised old waitress as Mount Columbia. What was there in a bare fortnight's grub supply upon which to base a siege of a mountain to which a million years was no more than the click of a focal-plane shutter?

I really intended to expostulate with Harmon and try to persuade him to do something in a propitiary way before it was too late. I was just beginning to expostulate, indeed, when poor little "Tip," carried down by the current, was trodden upon by one of the pack horses, and we had to give our attention to his crushed and bleeding paw.

Early in the afternoon we reached what appeared to have been the base camp of the mountain-climbing party. With three or four miles of open flat still running on to where the river headed in glaciers under Columbia and King Edward, we were puzzled at first as to why camp should have been made so far away. When "Soapy" came back from a scouting canter over the flats and along the mountainsides, the reason was made plain. Lack of forage farther up the narrowing valley accounted for the location of the camp where it was. It had been pitched literally at "last grass."

It was a comfortable spot, well sheltered from the winds from the icefield by a thickly timbered island curving around it to the west and south. Wood was at the door of the tepee, with water close at hand both above and below. Two trees, very favourably placed as to direction and unusually high for so near timberline, were ideal for the radio aerial.

Added to these utilitarian advantages was the promise of rare beauty of setting—when the clouds lifted. And that—since it was not the stormy time of year—would be only a matter of two or three days at the outside. Harmon—and all the rest of us, for that matter—had no apprehension at all on the weather score.

Then followed eight days of futile waiting—an interval in some ways more trying than our worst spells of travel in mud and water. We were vouchsafed one unforgettable view of the slender pinnacle of the mountain we had come so far to photograph, with the snowy summit suffused in the golden-pink glow of the sun that set on the day of our arrival. Then the most beautiful peak of the whole North American Rockies system settled down to the provocative tactics which had made it a mountain of mystery since the day of its discovery.

One day it was a chaste madonna, raising an adoring face behind a dusky veil that barely revealed a misty outline of its form. Another it was a dancer of fire and verve, lifting a tantalizing skirt to show dimpling knees twinkling in the froth of billowing chiffon, or poking a coquettishly bared shoulder from behind a masking screen. A third day, with an opaque stratum of cloud cutting off all but its upper and lower extremes, it leered brazenly after the manner of the chorus girl of a burlesque dancing in a barrel.

This was all good movie stuff, as far as it went; but it was the classic "altogether" that we wanted for the permanent record of the stills.

The half-famished horses were munching willow-bark and leaves in place of the grass already gnawed to the roots. Our own salt, sugar and canned goods were entirely gone, with only a much-reduced ration of bacon, flour and musty oatmeal remaining. Still we hung on, waiting for our perverse minx of the mountain to exhaust her whimseys and, as "Soapy" put it, "give us an honest-to-goodness look-see."

By way of reward for our patience, what should she do but take the veil completely? With a four-day blizzard from the north reducing the width of our world to a bare 50 feet from the tepee door, we went right on waiting, cheered by the wonder of lighting which we told ourselves simply *had* to come when the proverbial sunshine followed the storm.

But we could not hide from ourselves the fact that old Mistress Columbia had picked up our lightly flung gauntlet and was giving us siege-for-siege. And *we* had only eight days' rations left on which to complete the siege and reach Jasper, while old Columbia, who needed only snow to live on anyhow, was getting more of it every minute!

It was in these last stormy days—the dark before the dawn, so to speak—that our Radiola furnished us with the greatest enjoyment and comfort. There in our wretched little camp on the Arctic side of the divide, with the snow banked two feet high against the windward side of the tepee and the roar from the blizzard-tossed trees rising at times to a crescendo that drowned our voices, that blessed little black box reached up and teased things out of the ether in a way we had never had them before, even at Athabasca Glacier.

The fact that they had to reach us across the whole 150 square miles of the Columbia Icefield, and over the tops of several of the highest peaks in the Canadian Rockies, seemed to make no difference at all. Or if these things did make a difference, it was by way of improvement. It is probable, however, that the remarkable radio reception we experienced at the head of the Athabasca was due more to the fact that, for the first and last time in the whole course of the trip, we had ample leisure in which to string up the aerial carefully and give the set a fair chance. With similar attention elsewhere, doubtless it would have performed just as well.

How the amazing little box could still go on with its tricks after the bangings and sousings it had received is one of the unsolved mysteries of the expedition, along with "Tip's" miraculous "second-sightedness" in the matter of topography hitherto unknown to him. There still appear to be a few things in heaven and earth undreamed of in the philosophy of Horatio.

One of our most interesting radio experiences at the head of the Athabasca was that of picking up the Wills-Firpo prizefight during daylight. It was not storming at the time, but the weather was cloudy, with a threat of snow at any moment. There was no question that it was snowing on all the higher peaks and the Columbia Icefield.

We did not even know that the fight was scheduled to take place, but had no trouble in deciding what was afoot on cutting into some such announcement as, "Wills stopped the 'Wild Bull of the Pampas' with a stiff uppercut to the jaw." When the round closed we learned that we had been hearing a blow-by-blow account of the fight, described from the ringside and rebroadcast by KFKX of Hastings.

That accounted for a wild, pulsating roar, running like an obbligato through the terse announcements of the results of each blow. It was not the howling of the intervening storm but only that of the excited mob in the amphitheatre.

We had cut in at about the fifth round, and had followed the milling to the seventh or eighth, when a new sort of roar began to drown that of the cheering fans. "Soapy" suggested it might be the sound of a knockout, adding that he wouldn't be too sure about it, as he had never heard one before. Yet a knockout it proved to be, though a somewhat different one than the wise old packer intended.

The mysterious roar was quickly localized, resolving itself into the crashing of underbrush, wild snortings and the excited yelpings of dogs in full cry. An instant later the pack box containing the Radiola and batteries was jerked from the pile of saddles upon which it had been set and dragged violently against the side of the tepee. The headphones were yanked from our ears.

Rushing outside, we were just in time to see a speeding bull caribou disappearing into the timber, with the dogs yapping at his heels. Why they had been able to keep so close to their quarry did not become quite clear until we discovered that the animal was running under a handicap—that of a couple of hundred feet of the trailing wire of our wrecked antenna.

What had happened was simple enough, once we had time to figure it out. The dogs had started the big bull in the flats to the west of our sheltering island. Dashing into the thick timber of the latter, the frightened animal had come crashing out into our camp, snagging the lead from the antenna on the prongs of his wide-spread antlers as he rushed by the tepee.

We found all of the looted wire inside of a quarter of a mile, but not in time to string it up again to catch any more of the blow-by-blow story of the Wills-Firpo argument.

Both KGO and CKCD gave us detailed accounts of the fight by rounds when they came in that night. It was agreed that Wills had all the best of it, though a good deal of awe and admiration appeared to have been aroused by the "devastating rushes of the 'Wild Bull of the Pampas.'" Which led "Soapy" to observe (and not without point, one must admit) that if the rushes of "Wild Bull of the Pampas" had anything on those of the "Wild Bull of the Athabasca," he couldn't figure where the nigger had a look-in. Then, as a possibly clarifying afterthought, he added: "Mebbe they just raise rushier bulls hereabouts."

Another memorable radio event was the listening-in on the broadcasting by KGO of a complete opera rehearsal from the Capital Theatre, San Francisco. It was the newly formed San Francisco Opera Company, about to inaugurate its autumn season at the Municipal Auditorium. Advised some days in advance of what was planned, we were all tuned in and waiting when the program started.

There was something strangely familiar in the clear, ringing voice of the special announcer of the evening, and it was not long before I felt sure I had recognized it as that of my old college friend of Stanford, Charley Field, now editor of *Sunset*. I had spent six weeks on a commission with Field in China some years previously, and so had some line on every trick in his oratorical repertoire. An interim announcement by the regular KGO man, thanking "Mr. Charles K. Field of *Sunset* Magazine and the Bohemian Club for his masterly aerial conducting," shortly confirmed my suspicions.

The solos and other numbers by the stars came in with great clearness. The singing of the chorus had a muffled, diffused sound that was possibly due to the fact that special provision had not been made for concentrating it before passing on to the microphone.

One little unrehearsed incident was rather amusing, even at a distance. Field, after introducing one of the stars as Madame something, was called sharply to account when the temperamental prima donna took the air herself to say it was a mistake to prefix Madame or any other foreign title to her name. She was just a simple American girl—just plain Bridget Maloney. Or was it Eikenstein, or Pirelli, or Delysia, or Vodkavich? In any event, it was just some plain, simple old Yankee name that there was no possibility of mistaking,

Charley Field is blessed with a rare and delicate sense of humour. If I didn't catch his chuckle after that sally it was because he put a hand over his twitching mouth and turned away from the microphone.

My own great personal thrill from the radio came the night one of my carrier pigeon letters was read back to me over the air. This had gone out by a bird released at Castleguard, addressed to Walter Woehlke, managing editor of *Sunset*, with some corrections for one of a series of articles he was running on my Colorado River voyage. It also mentioned the names of several stations we had just heard over the radio.

Arriving at its bearer's home cote in Banff, the wrinkled wisp of paper with my message typed upon was it forwarded on to its destination by mail. Woehlke promptly handed it to his colleague, Joseph Jackson, who was just leaving for Oakland to give one of his regular literary talks as a part of KGO's Monday night "Educational Program."

Jackson, when his turn came, told his hearers in brief outline of our mountain expedition, and read them my letter. Then, expressing the hope that I was listening-in, he told me the corrections I desired had been made. Finally, he gave a very kindly review of my *Down the Grand Canyon*, the first news I had that the book was off the press some weeks earlier than originally scheduled.

But diverting as it was—and as practically useful in keeping our minds from dwelling too much upon what finally appeared to be the complete failure of our attempt to photograph Mount Columbia from the north—neither the horses nor ourselves could eat the radio. The plight of the pack animals was even more serious than our own. We had plenty of fresh meat, and something less than a half ration of flour, bacon and coffee. The horses had been starving for six days so far as grass was concerned, with leaves and bark not doing much more than keeping them alive.

When, on the evening of the seventh day of our futile vigil, the weather bulletins broadcast by the Pacific Coast radio stations forecast another general storm moving in from the ocean, we had to confess ourselves beaten. "Soapy," saying that the hungry horses were likely to bolt for the lower valley at any moment, announced that we would have to leave the following day if we wanted to have the use of pack and saddle animals.

Harmon, who had been the most hopeful member of the party from the outset, threw up the sponge in the morning when the sunshine, passing by the still cloud-veiled mountain peak, came only to spatter the snowy valley with a spurious golden flood. Packing his cameras in great dejection, he told "Soapy" to get the pack train under way as soon as possible. It took some argument on my part to persuade him that no time would be lost if the two of us, letting the men and horses go on, remained to press the siege a few hours longer.

With the gaunt, hollow-eyed horses barely able to totter under the depleted loads, the pack train set off down the valley at noon of September 25. In spite of the fact the sky was overcast and threatening another storm, Harmon and I, with our saddle animals and the horse packing the cameras, remained behind on the off chance of the altogether improbable clear-up. Pushing up the valley at a leisurely pace, we reached a sheltered point two miles from the base of the mountain, ate our lunch and waited for something to turn up.

At 3:30, with a grey barrier of mist still masking the mountains to the southward, it began to look as if our only diversion was going to come from a belligerent bull caribou, who made a series of short, broken rushes in our direction by way of showing his prowess to a small bunch of palpably admiring cows.

Then, suddenly and without warning, the veiling clouds fell away like a parted curtain and Mistress Columbia, garbed in a clinging mantle of new-fallen snow, radiant in the calcium-like glow of the low but brilliant afternoon sun, stood bowing, "At your Service!"

Both light and setting were beyond anything we had dared hope for—sparkling side-shafts of sunshine, with just enough clouds for background and shadows.

It lasted for just 40 minutes—ever changing but ever beautiful—and in that time we exposed still negatives at the rate of one a minute, besides running 400 feet of film through the movie cameras. The black rectangles of paper torn from Harmon's film packs were piled up behind his tripods like the brass shells around a hard-pumped machine gun at the end of a battle.

And a battle this had been, in a sense—a battle in which, after tasting all the bitterness of defeat, we had snatched a golden victory at the last moment. That, as I think of it now, was the high moment of the trip.

The sooty pall of nimbus which rolled down from the north to snuff out the radiance streaming over Columbia pelted us with a spatter of snowflakes. Packing up the cameras, we rode out of the timber 15 minutes later into the teeth of a nascent baby blizzard.

Fortunately—with heavy mittens, fur caps and parkas—we were well prepared for such an onslaught and so did not have to seek shelter. The horses were a bit at sea until they struck the trail of their mates; then they bent their heads to the gale and plugged doggedly on down the valley. With the river at a low stage from the cold, the fords were easy.

The storm ceased just before dark, giving us rather a better chance for the last mile into camp through fallen timber.

On one of the three days which we camped at the mouth of the Chaba, to give the horses a chance to graze and recover their strength, we took the opportunity to make a hurried side trip to historic Fortress Lake. This lovely body of water, which drains both ways from its seat on the Continental Divide, was known to the Hudson's Bay voyageurs and frequently trapped in later years. In the present century it has probably not averaged the visit of an outfit a year.

The timber on the British Columbia side was finer than any other we had seen in the Rockies.

Leaving the mouth of the Chaba on September 29 with light packs, we reached Jasper on the afternoon of October 1. The Athabasca was followed fairly closely all the way, the last two days over a well cut-out trail. Below the mouth of the Whirlpool we were on the old transcontinental trail of the Hudson's Bay traders.

Some of the cuttings we examined may well have been stumps from one of the very earliest clearings. They looked fully as old as a number of rotting stumps of great size—doubtless left by the same axes—which had thrilled me in passing the site of Boat Encampment, at the apex of the Big Bend of the Columbia, four years previously.

Chapter XI
BACK THROUGH THE SNOWS TO BANFF

WE WERE EATING LITTLE BUT dried fruit and lumpy "doughgods"—mixed without baking powder or salt—the last two or three days down the Athabasca. After this frugal diet, it was with a zest which only a man who has been underfed in the open for weeks can understand, that we sat down to a feast prepared to celebrate our arrival in Jasper. The Jackmans, old friends and trail companions of Harmon, were sponsors of the felicitous affair, which they had been planning ever since word had come that we were shortly to arrive from the south.

I shall not attempt to list what the trail-hungry pair of us consumed, but I remember three helpings of lamb and mint sauce, and at least an equal number of ice cream and a wonderful thing, imported from Vancouver but called a "Boston Cream Cake." The quivering whorl of that whipped-cream pinnacle I shall remember in steel-sharp relief when the vision of the snowy crest of Mount Columbia has faded to the dimness of ancient tapestry.

The successful photographs of Mount Columbia and the head of the Athabasca completed the picture program as originally laid out, leaving the return of the pack train to Banff as the main problem to be considered.

With the winter's snows already lying deep in the higher valleys and passes, no time was lost in Jasper. We attended to the shipping of exposed film the afternoon of our arrival, and to reprovisioning and the purchase of heavier winter clothes the following morning. That left us free to depart early in the afternoon of October 2. The men and the horses cantered off at one o'clock, leaving Harmon and me to finish our mail and follow on as convenient.

Just as we rode out of town the special train of the Prince of Wales pulled into the station, the party entering waiting autos at once for the ride to quarters prepared for them in Jasper Lodge. As we were passing the Lodge an hour later, a friend of Harmon, stopping to shake hands, told him that the royal guest had gone for a walk down the road we were following, and that we might expect to meet him returning.

This was interesting news. I had met the Prince twice previously— once at his regimental mess at the Front in France, and later aboard my ship at Rossyth. This would be a fine chance to renew old acquaintance, I told myself, and especially to learn something of the way things had gone at the international polo matches at Meadowbrook. The Prince had been there in person, while we had nothing but a brief account over the radio. One was not constrained to meticulous observance of court etiquette in the northern Rockies, I assured Harmon confidently as we trotted along beside the winding forest road.

I have no doubt that we would have had a very nice little yarn there in the forest—except for one thing. I had failed to reckon with my six weeks' growth of whiskers, which had spread to proportions positively Bolshevikian. I understood better what happened when the radio informed us, a night or two later, that there had been renewed rumours of plots against the life of the Prince of Wales and that he was being guarded more closely as a consequence.

Presently the rounding of a bend of the road revealed the royal hiking party close at hand. The Prince, garbed in golf togs, was striding with energy a pace in advance of three officers in uniform, one of whom had an empty sleeve.

Two of the officers instantly clapped hands to side-pockets and came on, boring me with concentrated frowns of suspicion. The Prince did not falter in his stride, but both Harmon and I recalled later that his jaw dropped perceptibly, as if the surprise of seeing the vicious apparition looming up ahead had not been entirely a pleasant one. Certainly there was no suggestion on his serious face of the frank, boyish smile I had rarely seen it without before in public.

Sensing at once that smock-clad individuals with Bolsheviki whiskers were distinctly *persona non grata* with at least the determined young men who had the safety of young Prince Charming in their capable hands, I kicked a heel into the ribs of the half-reined-in "La Belle" and sent her ahead at a trot. Incapable of registering "Innocence of Intention" with the only phiz available at the moment, I did the best I could with my hands. I did not quite go to the length of raising them above my head as we met and passed, but both of them were conspicuously in sight. Nor was either of them near a side-pocket, which was more than at least two of the other group could say.

I am inclined to think I was the only one of the sextette to get any real kick out of the episode. The Prince's nod of greeting was curt and rather nervous. Two of the officers did not nod the breadth of an eyelash. Harmon was too offended, for the moment, by the coldness of the royal greeting to see the funny side of it. Even my own whisker-bristling grin was slightly forced. Frowning young men with hands in their pistol pockets are just a bit destructive of aplomb and *savoir faire*.

We found our camp pitched for the night at Maligne River, near where the auto road comes to an end. The next day we ascended along the river to the outlet of Medicine Lake, which we skirted over a snowy trail to camp at its upper end. This strange body of water, owing apparently to the fact that its bottom is composed of broken rocks, is bank-full only during the season of spring thaws. The rest of the year, with the underground drainage of greater volume than the inflows, it sinks to a level 50 feet or more below its high mark of the spring. During most of the year nearly all the flow of lower Maligne River runs by subterranean channels from the bottom of Medicine Lake.

Half a day from Medicine Lake brought us out to the meadows at the lower end of Maligne. Here, both to rest the horses for the hard work in prospect over the snowy passes and to photograph one of the most beautiful lakes of the Canadian Rockies, we remained three days.

With the tepee erected in one of the most picturesque settings of the whole trip, we took the occasion to make a number of movie camp-shots, long deferred for want of the elusive synchronization of favourable weather and location. The esoteric mysteries of setting up the tepee itself were shown in one of these shots; in others, the shoeing and packing of horses and the inner workings of the camp cuisine.

As Maligne is notorious for its stormy weather, we were not very sanguine on the score of having really first rate lighting for scenic shots so late in the year. But the lucky star which had served us so well at Castleguard, Saskatchewan Glacier and Mount Columbia again interposed in our behalf. The brilliant side-lighting on glaciers and mountain walls during the brief hour we had at the head of the lake, between going and returning canoe voyages, was all and more than one was justified in expecting later than July or early August.

The view up the lake from a high point half a mile above the Narrows, with a slender timbered peninsula in the foreground, the sparkling emerald waters in the middle distance, and snow-crowned mountain peaks and the blue-green ice of hanging glaciers reared against a vault of sapphire sky for a background, is one of the most perfect settings of its kind on the continent.

It was at our Maligne Lake camp that the radio brought in the result of the first game of the World Series, won by New York after 12 innings. We were not set up early enough to get the play-by-play broadcasts, but many stations—including several Canadian—were willing to tell us all about it in the evening.

We gave the radio several trials by daylight before leaving Maligne Lake, which was the last point on the trip where there was any chance of having time to spare for much beyond the regular grind of routine. We heard parts of the speeches of General Pershing and Secretary of War Weeks, in connection I believe, with the Defense Day program. We also had portions of Secretary Hoover's address to the Radio Congress in Washington, broadcast from the telephone by KGO.

We heard the strong station KFKX of Hastings very frequently during the later weeks of the journey. A change to a wavelength of 291 metres, which we had heard announced while in camp at the head of the Athabasca, seemed to help us greatly in picking up and holding KFKX. The genial announcer, Bill Hays, "with Mrs. Bill at the piano," was always welcome with his basso profundo solos, such as "Asleep in the Deep" and "Sailor Beware!"

There was diversion, too in the little local touches from Nebraska, such as the reports of corn-husking contests, or a plan of one of the railways to improve the quality of stock by trading the farmers' new bulls for old. But when we cut into the middle of a KFKX program on three different occasions, each time to find a learned woman from the state university giving one of a series of six (or perhaps it was 16) discourses on Parliamentary Law, it was too much. When this disaster was visited upon us for the third time, we simply laid the headphones down on the sounding-box of the pack case and let the dogs harry the disturber, as they had done with the lady poetess from Seattle with a name like "Carry-me-homah."

Kipling was right. Some things are "just too cruel hard to bear."

We broke camp early on the morning of October 7, skirted the lower end of the lake, and began the long but gradual climb along upper Maligne River toward our first pass. Snow began falling a little after noon, increasing by several inches the half-foot already on the ground where we halted to make camp. We had already left far behind us the last bare earth—save for windswept cliffs and small patches under thick trees—we were to see for many days.

The horses, rested and strengthened by the rest and prime grazing afforded by the three days at Maligne Lake, were in fine fettle and condition. The moment their packs were off and the inevitable back-scratching rolls over, they started a purposeful pawing for grass which augured encouragingly for their ability to forage in the deeper snows ahead.

It was snowing again at daybreak but began clearing before breakfast was over. The cold seemed to be increasing, however. A movie which Harmon attempted to take of the horses being driven into their rope corral before packing was spoiled by the repeated condensing of moisture on the lens. After that we began leaving the cameras outside of the tepee at night in order to keep them at the same temperature as the air in which they were to be operated.

There was a foot of snow on the trail all the way to timberline, two miles above camp, and almost twice as much when we reached the summit of Maligne Pass. This made slow, tiring going, though the slope was not considerable on either side of the divide.

A landscape of glaring, unbroken white proved very trying on the eyes, especially on those of "Soapy" and La Casse, who alternated in the lead. The man breaking the trail had to keep his gaze continually roving over the snow for possible obstacles; the rest of us could cut down the exposure to the glare with close-pulled caps or by keeping the eyes on the horse's head. There were supposed to be snow glasses somewhere in the outfit but no one had thought of digging them out.

The way down from Maligne Pass was by a very rough tributary of the Poboktan, which the men called Goat Creek. Snowdrifts and rocks, with little chance to find such trail as there may have been, turned the descent into a sort of skidding scramble that played havoc with pack lashings.

We came down to the main Poboktan at a point about 11 miles above where that stream empties into the Sunwapta, a junction we had passed on our way down from Wilcox Pass. We now encountered the same trail and broken-down telephone line we had followed down the Sunwapta. They continued up the valley and over a divide to Brazeau Lake.

We could expect to have the benefit of the cut-out trail for only a few miles—unless we failed to get over the first high pass, of course. In that event there would be nothing left to do but work out toward the plains by the easiest routes available.

The stream of the Poboktan emerged from a closely boxed canyon not far above where we came down to it from Maligne Pass. The avoidance of the gorge necessitated a long, steep climb up an icy slope to the bench above. Here it was that the first effects of the hard foraging conditions imposed by the snows at the camp of the previous night were in evidence. The horses were blowing and tottering at the end of every 40 or 50 yards. It was possible to ease the saddle animals by ourselves walking; for the pack horses all that could be done was to rest them at increasingly frequent intervals.

Four miles of painful progress across a bleak, windswept plateau brought us to the valley of a little side stream. It was earlier than we had intended to stop, but the condition of the horses was always the controlling consideration. In this instance, with perfectly good intentions, we did them a disservice in halting at a point where there was very little grass to be found even when questing nose and paw had cleared a way to the earth.

Bewildered by the elusiveness of forage, the tired and hungry animals scattered widely during the night. It was one o'clock of the following afternoon before the last of them was rounded up again, and nearly three when we were ready to take the trail. Under the circumstances, five miles up the valley to the edge of timberline was the very best we could do.

Camp was made in the last patch of scrub pines at a point which we had come to regard as "the parting of the ways." For here a definite decision as to whether or not we should attempt to break our way back across the high passes, already firmly in the grip of winter, could no longer be deferred.

Continuing by the trail we had been following up the Poboktan, a low and easy divide would be crossed to Brazeau, whence it was easy to work on to the plains by comparatively low valleys, with improving grazing all the way.

Clambering up a low ridge above the camp, on the other hand, we were face to face with the steep mountain wall, of 2,000 feet or more, the drifting snows of which had to be surmounted before reaching another valley and pushing on to a vantage from which to tackle a still rougher and more forbidding climb.

With a mile or more of both passes—the snowy and the open

one—in full view from the ridge above camp, the issue was as clean cut as it was unavoidable.

Harmon was still extremely anxious to attempt to make what he had come to call his "snow-picture." There had, of course, been snow and ice in practically every shot made from the outset. But this was to be a picture in which there was to be nothing but snow—with a few incidental horses and men. Neither movies nor stills had ever been made in the Rockies of a pack train trying to travel in such snows as had already closed down upon the higher passes, and the opportunity to make such pictures was too good to pass up without a fight. Harmon was all for the snows.

"Soapy," personally—both for the sake of his rheumatism and his horses—would have preferred the easier way. His agreement with Harmon, however—albeit only a verbal one—was quite explicit in the matter of attempting the traverse by the high passes. In the same game spirit, therefore, that he had gone through with the crossing of the Saskatchewan Glacier and the long vigil under Mount Columbia, the old packer now acquiesced in the plan to face the snows.

"I'm willin' to buck ahead till the horses quit," he said simply, and that settled the matter for the present.

That was our coldest camp by all of ten or 15 degrees. There were some evidences of the place having been used before, but if any tepee poles had been left, there was no way of finding them save by explorative tunnels and shafts in the snow. As it was too late to start mining operations, there was nothing to be done except make the best camp possible without the tepee.

Harmon swung his diminutive "pup" tent under a scraggly balsam. Prolonged search finally uncovered enough poles to give the cook tent precarious support, and this was turned over to me to sleep in. The packers, building a roaring fire and adding all of the saddle blankets to their bedrolls, "chipmunked" under a wind-stunted spruce.

Fortunately there was no wind. No thermometer had been included in the outfit, but two traditional tests of the old Alaska "sourdough" gave some line on the depth to which the mercury would have declined. At midnight the cans of unsweetened condensed milk were frozen solid. That meant a temperature of a bit below zero. At four in the morning the contents of an opened can of sweetened condensed milk could not be pierced with the point of the blade of a hunting knife. That indicated from ten to 15 below, with the temperature probably continuing to fall until daylight.

As the cook fire in front of the tent, which I fed liberally at frequent intervals during the night, must have raised the temperature of the interior by several degrees, I am inclined to believe that a thermometer, hung up on a tree outside, would have registered from 20 to 30 below zero. In spite of this little touch of the Arctic, no one complained of having made an especially uncomfortable night of it. Certainly we experienced several unhappier ones, notably on occasions when storms blew the smoke and embers of the fire about the tepee.

The horses were not hard to find in the morning. Unable to locate grass under the deep snow blanket, they had straggled down along the creek and done the best they could on willow bark. It was a terribly lean and gaunt-looking bunch that was rounded up in the rope corral, but most of the units of it still had spirit enough to go through the routine tricks by way of working off the night's accumulation of cussedness.

"Jerry," the movie horse, blew himself up like a toy balloon so that there would be slack in his cinches after deflation. He had done the same thing every morning he was packed since the beginning of the trip, but without ever seeming to learn that all it brought him was the rough prodding of Baptie's high-heeled boot, braced against the puffer's ribs in bringing the lash-rope to proper tautness.

"Wolverine" went on the air in his regular daily endeavour to broadcast the radio set by standing on his forelegs and kicking the hind ones. "Roan" trying to crook a loving leg around "Soapy's" neck, and "Buckskin" lying down and feigning cramps, were vigorous recrudescences of old stuff that gave some assurances that something more than the bare spark of life could be kept flickering on willow bark.

Saddling up at once, Harmon and La Casse, with the movie outfit, went on two hours ahead of the main outfit. This was primarily for the purpose of attaining a favourable vantage from which to make long distance shots of the pack train as it came winding up the pass, but it also gave opportunity for breaking trail unhampered by the crowding of horses pressing up from behind.

The whole route to the summit of the pass was in full view as soon as we had worked up through the timber and come out on the crest of the ridge above. Greatly foreshortened, with nothing to break the smooth expanse of the snow and give perspective, the ascent appeared even steeper than it actually was. The final thousand feet loomed as a veritable wall.

Through the high, thin air the distance to the crest of the divide seemed hardly a pistol shot. How considerable it was, with a hint of the difficulties of the going, became evident the instant our squinting eyes discovered Harmon and La Casse just beginning to ascend the culminating snow wall. After two hours of floundering through the snow, they had covered but two-thirds of the ascent, with much the hardest part still remaining to be surmounted.

Our own progress was made much easier by having only to follow a trail already beaten down by the feet of three horses, two men and one dog. Going also improved greatly the farther back one was located in the line of the pack train. In my enviable station at the rear of the labouring outfit it was really not so bad—for a little while.

As Harmon and La Casse commenced the ascent of the final pitch, their figures were silhouetted against the white snow with the sharpness of that of a soaring aeroplane against the sky. The horses looked no bigger than flies crawling on a ceiling, with the other figures proportionately more minute; yet queer, fluttering movements of the sextette of marionettes told the whole story of their desperately bitter struggle.

A twinkling flea in front of a swaying fly was La Casse, breaking trail with his own stout legs and leading his saddle horse. The flea in the rear of two other flies was Harmon prodding on the remaining horses. A restless, jumping pinpoint was "Buster," gambolling in the deep snow and having the time of his life as usual.

A sudden skidding of two of the flies indicated a plunge down the mountainside of as many of the horses. As we discovered from the marks later, the slide had been all of 40 feet. To us, the double furrow of ploughed snow appeared about as two-inch-long lines on a sheet of paper.

The struggle to get the rolled horses back on their feet was written as clear as skywriting. One who ran could have read it without skipping a line—one who ran *downhill*, I mean to say. To one who laboured up, especially if encumbered with the something like 40 pounds of clothes with which he had fought off the frigid temperatures of the night before, reading—even light reading—was not so easy. I may as well confess that, for the next three hours, I was a lot more concerned with, and deal sorrier for, my own puffing, floundering self than for any other of the score or more animate units of the outfit.

To the foot of the steeper incline to the summit it was just hard, steady plugging, with increasingly frequent rests to allow horse and man to recover breath. Then, by way of a real warming up for the final effort, the only possible route turned and led across what appeared to be a 300-yard-wide slide of broken rock slabs of great size, covered with from three to five feet of light snow. To the labour of ploughing through the latter was added the danger of broken legs for the horses, an ever-imminent risk among big rocks.

The three horses with Harmon and La Casse had evidently got out of control here, each plunging through on a wobbly furrow of his own. Not to be outdone, the horses of the main outfit opened out in a wide fan and added a dozen more individual trails to those of the leading trio. It was all of half an hour before we had them bunched again on the farther side.

"La Belle" and I had a serious dispute here over the right of prior occupancy of a cosy little foot-wide crack between two very unstable-minded slabs of limestone. Not having as many clothes on as I did, the spirited young lady, whom I was leading by her bridle, kept pressing impatiently upon my lagging footsteps. It was an annoying way she had on hills, though it never troubled seriously unless I was trying to trip lightly from rock to rock, or from bog to bog. As long as I had my wits about me, it was usually easy to keep her at arm's length with a back-extended hand.

Blowing like a freshly hooked grampus, and with my mind engrossed with the complicated navigational problems unfolding at every step across the slide, I neglected to give "Belle's" oncoming nose the straight-arm when skidding hob-nails set me sliding into the grip of a rocky pair of yawning jaws. As a consequence the mare's off forehoof drove down into the hole about a hundredth of a second after it had been occupied by my near hind one.

The rocks—or at least one of them—were as much agitated about the difficulty as was "Belle" or myself. The smaller of them lost its balance, but, contrary to what usually happens in a case of the kind, teetered outward rather than inward. That relieved the jam and allowed me to withdraw my leg and let the mare have the hole to herself.

Her sharp-shod hoof had been planted solidly against the side of the toe of my big rubber overshoe, but three or four pairs of heavy woollen socks effectually buffeted the pressure. The rasp of a bridle buckle across a whiskered cheek left more of a sting in its wake than the bump from a hoof which had been planted a half inch too far to one side to do any real harm.

The final thousand feet was just a bit more than hard work, for there were two or three places where the slope was so sharp that one misstep meant a rolled horse and a scattered pack. Helped greatly here by the path already broken ahead, we had no serious trouble.

Our worst difficulty on this part of the climb, indeed, came from breaking through the crust of an earlier snowfall—probably one put down by the storms which had assailed us at the head of the Athabasca. The three feet of new snow on top of this crust was enough to carry the horses only if they moved steadily ahead. The least bit of floundering put them down into the older snow, with the sharp edges of the crust gashing their legs. Once through, it frequently took many yards of painful wallowing before getting back on top of the crust again.

With 500 feet more to go, we touched the edge of the widening swath of sunshine which was slowly rolling back the pall of shadow that had enfolded the ravine by which we had ascended. In the strange way it has in the rarefied air of high altitudes, when there is no wind blowing, the sunlight stabbed like a spurt of flame. That sparkling golden shaft of light was more than hot—it was burning to the skin, blinding to the eyes.

My abused anatomy, toiling in the heart of the ponderous pile of wrappings which had protected it from the Arctic rigours of the night, had been developing the temperature of a Turkish bath even in the shadows. With the touch of that withering wave of sunlight the genial warmth of the Hammaam was converted to the scorching heat of the furnace blast.

Regardless of my hitherto jealously maintained place at the rear of the line, I checked "La Belle" and commenced a swift stripping off of superfluous garments. A shooting jacket had followed a 15-pound duffle coat, and a sweater and lumberjack's shirt were next in order for the bundle to be lashed to the pommel of "Belle's" saddle, when a far-carried shout assailed my ears from the crest of the pass. It was Harmon, who had been watching my disrobing act with his glass, yelling through cupped hands.

"Leave 'em all on!" he was shouting. "Look better for movie!"

Realizing that I ought to humour the veteran after all his work in helping to break out the trail, I called him a few lurid names and then did as he asked. Almost down on my hands and knees at the finish, I dragged my reeling bulk up to the summit a good 200 yards astern of the last of the horses. The interval was so great that Harmon had to make a separate shot of my arrival.

Turning "La Belle" over to the packers, I slunk off and slithered down in the snow. Presently Harmon came over and began speaking.

"Sorry to have troubled you to drag all those togs up that last stretch on your back," he said. "Great waste of energy; also of film. It wouldn't do, you know, to run a shot of a man bundled up in that North Pole rig-out just after "Soapy" and Rob had reached the top in their shirt sleeves!"

"Wow!" I had the wind for just one demoniac yell; but faltering flesh denied the demand of outraged spirit that I run Harmon down and beat him into insensibility.

Resting for half an hour on the warm, windless summit to breathe the horses and relash the packs, we began the descent of the southern side of the pass to the broad, open valley of Jonas Creek. There was not much to choose between one slope and the other. Neither side had a trail, or, if it had, there was no finding it. Snow was drifted deeper on the Jonas slope, but there was no place where the horses could roll far without regaining their feet, and no deathtrap of a rockslide to cross.

Several times horses which strayed from the line of descent carefully broken out by "Soapy," dropped over low ledges to disappear from sight in the drifts below. After a billowing and heaving of the snowy blanket, a wobbly hump would begin to run through the soft mass, and presently a horse came wallowing out below, not any the worse for the experience.

Harmon was almost in tears because these falls, with their subsequent tunnellings, kept eluding a movie shot. Old "Soapy," however, already beginning to feel his first touch of snow-blindness, sternly refused to delay progress by staging the action. We were over the pass, he said, but had only made a quarter of the distance that would have to be covered before reaching a camp in the timber. It was no time for frills. We were still well above the timber even after dropping down 1,500 feet or more to Jonas Creek. Here we began a gradual ascent of several miles which carried us over a hardly perceptible divide to Ram Creek, running down to the Brazeau on the east. Deep drifts were still troublesome here, but the worst trial was that of the glare.

Search through the packs for snow glasses at the previous

camp had revealed that these almost indispensable protectors had been sent off by mistake at Jasper. "Soapy's" eyes gave out in the first mile up Jonas Creek, Baptie taking his place at breaking trail at the head of the line. When the glare-dazzled snows had inflamed the wrangler's eyes to such a degree that they could no longer judge fall and distance, La Casse took the lead and held it on down Ram Creek to the timber.

Harmon and I felt the glare badly as long as we continued taking pictures, but the irritation was eased a good deal when we returned to line and reduced the dazzling landscape to a narrow peephole by pulling down "Balaclavas" or puckering the strings of parka hoods. It was the trail-breaker, forced to expose and use his eyes every moment, who had the hardest time.

Our camp on the Brazeau, just above the mouth of the Ram, was the most comfortable we had pitched since leaving Maligne Lake. It was on a sheltered bench, a thousand feet below timberline. There was still much snow, but on the level river flats this had blown thin in places, so that the horses lost little time in pawing through to the grass.

Because of having been frozen and buried in the snow while it was still green and uncured, the grass was hardly as nourishing as it looked. It was so much better than the horses had enjoyed for a long time, however, that "Soapy" was anxious to remain over a day or two to give them a good feed of it before tackling the next high pass, now looming high above us to the south.

"Soapy's" eyes were in the worst plight from the snows, but even he did not have a spell of complete blindness. None of us had a comfortable night, however, and all were complaining of the sensitiveness of their eyes to glare during the remainder of the trip. Fortunately, most of the rest of our traverses above timberline were made in cloudy weather.

The irritation of the membranes of the eyes incident to what is commonly called snow-blindness, was graphically described by an old Alaskan "sourdough" friend of mine as "like a handful of red-hot sand chucked under your lids."

The horses looked so much better, after their day of rest and improved grazing, that we had about decided to give them another one, when the radio brought in word which seemed to make further delay out of the question. A general storm which had broken upon the northern Pacific coast, KGO informed us, was expected to continue for several days.

Judging by the experience we had already had with storms the radio had told us were working eastward from the ocean, this meant that the moisture-laden clouds would be condensing upon the high peaks of the Continental Divide during the night, and that we would find it snowing in the morning. Beating the storm across the next pass—the worst one remaining to be surmounted—before it developed to blizzard proportions was our only chance to continue through the high mountains and complete Harmon's "snow-picture."

A gusty northwest wind began scattering the tepee fire a little after midnight. Before morning it was blowing hard and purposefully, accompanied by flurries of snow, dry and powdery enough to have come from the Arctic. The clouds themselves were high, but already the peaks flanking the pass were obscured in clouds of wind-driven snow.

"Soapy" shook his head dubiously, as he blinked through goggled eyes at the fluttering sheets of whiteness masking the notch of the pass, and reckoned we would be scurrying back to the valley before noonday to thaw out our ears. Just the same, he added, he was game to go on with the geezly stunt—quite ready to stick it out as long as the cayuses would. Five minutes later he was leading the pack train across the windswept flats to the mouth of Ewe Creek, by which the ascent was to be made.

For a mile we toiled up through close-growing timber, where the wind reached us only in spinning gusts. Not until we came out to an open plateau above the last of the trees did we feel the full force of the gathering storm. The first blow of it seemed almost as solid as the slap from the side of a board, and one had to close his mouth and breathe slowly through contracted nostrils to keep the flying ice particles out of his lungs.

For a few minutes—until the more pressing need of things made demand upon mind as well as body—a verse of "Screw-Guns" kept running through my head.

> The eagles is screamin' around us, the river's a-moanin' below,
> We're clear o' the pine an' the oak-scrub, we're out on the rocks an' the snow,
> An' the wind is as thin as a whip-lash what carries away to the plains
> The rattle an' stamp o' the lead-mules—the jinglety-jink o' the chains—

For a half-mile, where the last aspiring tongue of stunted trees ran out in wind-flattened bushes, we had to watch closely for the snow-covered hummocks to keep from floundering onto and breaking through them. Then these were left behind, the last of the jutting rock pinnacles were drifted over, and we tunnelled upward in a universe of unbroken whiteness.

Just how much of the snow which assailed us as we struggled up between the converging walls of the pass came direct from the clouds it was impossible to tell. It was snowing, and snowing hard; yet there is little doubt that nine-tenths of the whirling blanket enfolding was made up of drift, blown from the peaks three and four thousand feet above. Borne by conflicting gusts, this struck us from all directions. When it came from directly ahead it was impossible to penetrate it with the eye. We could only blink, owl-like, until the next gust brought the blinding cloud behind us and made it possible to open the eyes again.

I had faced heavy winds in many of the great passes of the Andes and Himalayas, but never a real storm. My outstanding memory of this occasion was the terrific roaring of the winds among crags and cliffs which were entirely cut off from sight by driven snow. It seemed impossible that a sound so deep and raucous could come from the friction of air on rock. Time and time again I reined in my horse in the fear that an avalanche was descending just ahead, only to find that I was shrinking from the threat of a bugaboo no more tangible than thin air. Yet I have seen whole mountainsides of rock go down with no deeper, more savage a roar.

"Soapy," Baptie and La Casse—on foot, of course—took turns at breaking trail. This was quite as hard work for the man who did it as on the steeper ascent to the Poboktan-Jonas Pass, but it was far less helpful to the men and horses following. This was because of the terrific rate of drift. If an interval of 50 yards opened up in the line, the trail for the bunch behind had to be broken out anew. Half a minute would completely obliterate a three-foot-deep trench beaten down by the passage of a dozen horses.

The horses were wonderful—more sensible and easier to control than I remember them at any other crisis of the trip. Anything less than the spirit they showed in bucking the drifts in that savage storm would have forced us to turn back a mile above timberline. As it was, slowly but fairly steadily, we worked them through to the shelter of a ledge which La Casse recalled as being almost immediately below the summit.

It looked as if we had won the fight—that all was over but the inevitable shouting which would accompany the long, easy drive down Cascade Creek on the other side. We even talked a little of the nice, warm, comfortable camp we were going to make in the thick timber of the flats of the Cline, which we ought to reach in three or four hours.

All of which made it a bit more of a shock when, with the horses well rested, we pushed out from below the sheltering ledge to find our way blocked by what appeared to be an impassable wall of snow. Ulus remembered the place as a very steep pitch of broken rock. Snow, drifting over the crest of the pass, had converted this into an almost sheer slope.

"Soapy," anxious to avoid discouraging the horses by unnecessary exertion, went at the task of sounding out the barrier coolly and methodically. At point after point we would tread out a trail on foot, only to have the first horse break through the crust below and go down over his ears.

As a last resort, "Soapy," leading an astonishingly agile and stout-hearted Indian pony called "Roan," worked along the base of the snow wall and attempted to reach the summit by zigzagging up the less deeply buried slope beyond. Here, with the help of Baptie and La Casse, he was finally successful.

Leaving "Roan" on the summit, the men came back and started the remainder of the pack train. All went well along the base of the snow wall, but when the first zigzag was reached the leading horses began pounding their way through a deeply covered crust which had supported the light-stepping "Roan." Dropped almost out of sight, several of the nervous animals started to wallow around in their tracks and start back down the slope into the line of horses bunched below.

With three of the animals beaten and ready to quit, the packers made extraordinary efforts to prevent the spirit of defeat from spreading among their mates. Once a pack train gives up a crucial struggle of the kind in which we were now engaged, there is nothing left to do but turn back. Doubtless that is what would have happened in the present instance had the quitters been allowed to swing back and demoralize the rest of the outfit.

Calling on Harmon and me to help them, the men brought the leading animals back on to the trail, and then—five men to a horse—we dragged them, one at a time, up to the windswept summit. After the first three or four had gone the route, the trail was beaten deep enough to make it possible for the others to go up without other than vocal help.

Harmon, unlimbering his movie camera, made a shot of the tail of the procession. It was a pity, of course, that at least a few feet of film could not have been cranked on some of those first sounding flounders.

It was a long, tedious descent to the Cline, but we made it just before dark, to pitch a camp no whit less comfortable and cosy than the one we had pictured in fancy under the snowy ledge below the summit of the pass. Lower than we had been for many days, there were spots in the flats where the grass was actually showing above the snow.

The storm did not follow us down to the valley, though we could see for two days where it was raging among the high peaks and passes. There is no possibility that we could have crossed the Brazeau-Cline divide a day later. Beyond all doubt, our radio had saved his long-dreamed-of "snow-picture" for Harmon.

We made an easy stage of it to beautiful Pinto Lake the day after crossing from the Brazeau. Here we left the snow for a while, not to be bothered with it again until we reached the head of the White-rabbit, beyond the Saskatchewan. There was snow at several passes between there and Banff, but not enough at any point to make serious trouble.

Once down to the Saskatchewan, at the mouth of the Cline, we had a well-travelled trail all the way south. More important still, the route led by valleys famous for the finest grazing in all the Rockies. With grass for our long-starved horses and Harmon's "snow-picture" complete, there was really very little left to worry about.

One night toward the end of our journey, the radio, in spite of the waning strength of batteries which had served all the time since leaving Lake Louise, picked up a station in Davenport, Iowa. A local seed company was sponsoring the program, which consisted very largely of the proud reading of telegrams attesting that they had been heard in several neighbouring counties, and even in Nebraska and Illinois.

I only wish there could have been some way we could have winged them word of how and where our funny-looking outfit was listening-in upon them.

With the horses picking up weight and strength all the way, we cantered into Banff on October 24, ten weeks after our departure from Lake Louise. Not one horse had been lost; not one had been permanently lamed, in the whole course of what was probably the roughest continuous pack-train journey made in the Rockies since the time of the pioneers.

It was a notable achievement for our packers.

ENDNOTES

Foreword

1. Lewis R. Freeman, *On the Roof of the Rockies*, Calgary: Rocky Mountain Books, 2009.
2. Ibid.
3. Ibid.
4. Ibid.
5. Ibid.
6. Ibid.
7. Ibid.
8. Ibid.
9. Ibid.
10. Ibid.
11. Ibid.
12. Ibid.
13. Ibid.
14. Ibid.
15. Ibid.
16. For information on this trail, see Emerson Sanford and Janice Sanford Beck, *Life of the Trail 4*, Calgary: Rocky Mountain Books, 2009.
17. Freeman.
18. Ibid.
19. Ibid.
20. Ibid.
21. Ibid.
22. Ibid.

Chapter III

1. See *Down the Columbia*.

Chapter IV

2. Just why a glacier that is formed of snow, the universal symbol of immaculate whiteness, should be named from a bird whose jetty plumage is popularly accepted as the "blackest of things black," I have never heard satisfactorily explained. The jewel of consistency, however, has never been set in the forehead of the god who inspires the nomenclature of natural features. The straggling tentacles of ice in question really do have a remarkable resemblance to the leg and talons of some giant fowl, but, being snowy in colour, it is at least open to argument that some such bird as the ptarmigan or the white leghorn should have been nomenclaturally honoured in preference to the crow.

Chapter VIII

3. *Alpine Journal*, Number 227.

THE MOUNTAIN CLASSICS COLLECTION

Rocky Mountain Books is proud to make classic works of mountain literature available again. Many of the books in the *Mountain Classics Collection* have been unavailable to the general reader for years. Each volume includes the original text and includes a contemporary introduction written by experts currently living or working in the mountain community.

978-1-897522-50-9
$19.95

978-1-897522-49-3
$19.95

978-1-897522-96-1
$22.95

978-1-897522-06-6
$19.95

978-1-897522-14-1
$19.95